Escape from Quantopia

Collective Insanity in Science and Society

Escape from Quantopia

Collective Insanity in Science and Society

Ted Dace

BOOKS

Winchester, UK
Washington, USA

First published by iff Books, 2014
iff Books is an imprint of John Hunt Publishing Ltd., Laurel House, Station Approach, Alresford, Hants, SO24 9JH, UK
office1@jhpbooks.net
www.johnhuntpublishing.com
www.iff-books.com

For distributor details and how to order please visit the 'Ordering' section on our website.

ISBN: 978 1 78279 610 7

A CIP catalogue record for this book is available from the British Library.

Design: Stuart Davies

Printed and bound by CPI Group (UK) Ltd, Croydon, CR0 4YY

CONTENTS

Introduction

Wake up, the world's on fire!
— Lawrence Ferlinghetti

From a distance a pair of headlights resembles the pointy wings of an angel. Seated on cushions two feet off the ground, we glide to our destinations. Could this be the heaven foretold in ancient texts? And when it all comes crashing down, belching out flames along the way, is that hell?

The name of this book popped into mind one day as I raced down a stairway to the exit of my workplace. In the wake of the 2008 financial meltdown, I'd signed up with a temp agency that placed me at a property and casualty company insuring many thousands of clients. Every morning, I entered my cubicle and logged on to a galaxy made of ones and zeroes, my task to sequentially locate two multi-digit numbers so as to navigate to the correct window where I could then type in two other multi-digit numbers.

Except when the system inexplicably hurled itself over a cliff, that was it eight hours a day.

Not so long ago your insurance policy was your community. If your house burned down, your neighbors helped you rebuild. May not have been perfect, but by and large it worked. What it did *not* do was provide a means of capital accumulation for owners of insurance companies.

In little more then a century we've descended from the warm smile of humanity to the ice cold calculations of Quantopia.

Valuing profit over quality, agribusiness giants routinely lace animal feed with antibiotics, breeding resistant bacteria that make us sick. Doctors and nurses, meanwhile, have become cogs in a medical machine that turns the "human body into nothing more than a bunch of billing codes for the insurance industry," as an ex-nurse recently put it.[1] Teachers aren't so much sharing knowledge gained from life study as transferring data from point A to point B, followed by quantitative measurement of how much was received. With little substantive disagreement among candidates striving to please big-money backers, politics has devolved into a spectacle of dueling percentages.

In place of neighbors and colleagues and extended family taking care of themselves and each other, we have "the economy," in which all real-world values are converted into dollar amounts. To obtain a needed item means exchanging money for it, and since most of us acquire money by selling our labor on the open market, we ourselves are subject to quantification. Those who can't earn enough to pay the bills are worthless and require credit from the worth-more, always glad to lend their surplus since they get it all back with interest. While the few gain even more on top of their already excessive pile, the many lose what little we have.

Whether you're picking through garbage, juggling three jobs or building flying robots for the terror and death industry, you do what you do because money, the ultimate quantitative abstraction, decrees it.

Life is increasingly privatized behind computer screens, our interactions digitally mediated in a faceless and voiceless world. In place of touch we have Touchpads. Subjecting prisoners to years of solitary confinement doesn't seem so jarring when it only amplifies what all of us undergo to one degree or another. As spontaneity and creativity give way to narrowly defined routine, we forget the irreducible qualities that generate meaning and fulfillment. As Martin Luther King put it, "The profit motive, when it is the sole basis of an economic system, encourages a cutthroat competition and selfish ambition that inspires men to be more concerned about making a living than making a life."[2]

If capitalism is the face of modernity, science is the brain. Originally known as philosophy, the drive for knowledge accumulation took an early mathematical turn. Pythagoras and Plato substituted the tangible world of the senses with geometry, reducing a triangle to a flawed reflection of the eternal ideal triangle which, along with the other Forms, gives order to the world. Substitute "Forms" with "laws of nature," and we have the metaphysics of modern science.

Not to be confused with actual physics, which continues to uncover astounding complexity and nuance in the world, *physicalism* traces the source of all things, even human beings, to transcendent law. When Stephen Hawking says "biological processes are governed by the laws of physics and chemistry and therefore are as determined as the orbits of the planets," he's setting aside real science and speaking instead as a citizen of Quantopia, where mathematical laws generate everything from atoms to thoughts.[3]

"'You,' your joys and your sorrows, your memories and your ambitions, your sense of personal identity and free will, are in fact no more than the behavior of a vast assembly of nerve cells and their associated molecules."[4] Though the author of this comment, Francis Crick, co-discovered the structure of DNA, his assertion is based not on empirical inquiry but the subordination of science to physicalist dogma, according to which brain activities — far from facilitating our passions and our power of reasoning — follow entirely from immutable laws of energy and mass.

Though science is commonly thought to be rooted in the observation of tangible events, materialism is only the shadow of mathematical idealism. What counts is not the particulars but the universal principles drawn from them. Recognizing only deterministic law and blindly obedient matter, Quantopia is a world without heart. The strictly mechanical interplay of atoms according to timeless equations leaves no room for life or love or free will or anything recognizably human.

To escape this prison of our own devising, we must backtrack to the beginning and start anew. Fortunately this is not so difficult, as the cosmic reset button is always being pressed in a process we call *time.*

Lost in abstraction, physicalism denies the reality of flowing presence. As we shall see, physics itself demonstrates no such thing, but experimental findings are typically interpreted according to age-old unconscious bias toward the static and timeless.

Fixed principles are a poor fit for a universe that evolves into life. By proposing that species are shaped in part by the adaptive organisms that comprise them, Charles Darwin led a failed revolt against Quantopia. Instead of passively reacting to mechanical forces, organisms respond intelligently to changing environmental conditions with new behaviors that lead, over many generations, to new bodily forms. But Darwin never satisfactorily explained how adaptations are inherited by progeny, and his theory was replaced by a physicalist alternative euphemistically labeled *neo-Darwinism,* which casts evolution as a blind mechanism of random genetic mutation coupled with environmental selection. Meanwhile Darwin's observations about nature were misapplied to economics, giving rise to *social Darwinism* and its pseudoscientific justification for merciless exploitation via the equally blind mechanism of the market.

Darwin's name has long been packaged with "neo" and "social," the first projecting a fantasy of technological authority into the nucleus of the cell, the second re-projecting it from nature back to industrial society. By prefixing Darwin, Quantopia effectively neutered him.

A quieter yet more profound assault on the citadel was launched by American philosopher Charles Sanders Peirce. A younger contemporary of Darwin, Peirce wondered if evolution applies to all naturally enduring forms. What if biological habit is only a special application of a general tendency of nature? What if we're just imagining these quantitative principles that preordain the forms of particles and people? Perhaps the universe began without law or plan and, like a hatchling, had to feel its way forward, spontaneously establishing forces and properties of matter and maintaining them out of "cosmic habit."[5] In place of changeless laws, the world is built up from habits evolved over time.

More recently physicist David Bohm proposed a theory of nature based on the *implicate order,* which not only clears up many conundrums of quantum mechanics but yields what he called "cosmic memory."[6] Fellow quantum pioneer Walter Elsasser observed that organisms resist simple physical analysis and cannot be explained without recourse to "holistic memory."[7] Contemporary physicist Lee Smolin chalks up nature's stability over time to the "principle of precedence."[8] Regardless of what we call it, nature's memory is the means by which intrinsic forms endure.

Behind commodity trading and molecular reactions lie principles of exchange. The question is whether these principles are disembodied, without space or time, or simply follow from repetition in a process of universal habituation. Whether they determine the course of events or merely reflect established habit, the equations look the same.

Bohm and Elsasser learned much from philosopher Henri Bergson and his theory of time and memory, as did Rupert Sheldrake, the biologist who not only resurrected Darwin by explaining how adaptations can indeed be inherited but provided a testable hypothesis of an evolving cosmos. Yet Sheldrake's reward for universalizing evolution was to be rendered taboo among his colleagues.

Widespread hostility to Sheldrake's theory of natural memory via *morphic resonance* provides the starting-point for a broad look at the ease with which unreasoning becomes ingrained as group-think. Morphic resonance explains not only stability of form in nature but why the dogmatic reaction of biologists to Sheldrake so quickly and comprehensively eliminated any alternative response. Uniformity of human thought is no different, in principle, than uniformity of helium atoms.

Watson and Crick in no way demonstrated the existence of a genetic blueprint. The encoding of a developmental program or recipe in DNA is a modern fairy tale, a science fiction. While genes determine peripheral traits that distinguish one individual from another within a given species, we cannot explain the general form shared by all its members without simultaneously tackling the question of mentality.

Far from a complex property of nervous tissue, the mind is the body's living history, the means by which past actions influence current decisions, whether genetic, cellular, muscular or cerebral. Our failure to understand memory and mind has kept us in the dark about not only our self-existence but also a slew of self-propagating social evils. Senseless taboos, blind obedience, paranoid hysterias, national narcissism, war, patriarchy, racism and class divide — all are perpetuated, in part, by nature's tendency to remember and repeat.

How can a world so deranged, so wrong on so many levels, be the product of human beings? We're not so bad, right?

Mechanistic bias blinds us to the fact that the flow of money demarcates the movements of a living thing. The more money to be made, the more it spends itself, spinning off a self-organized system of production and waste with the moral intelligence of a hurricane. In a society that rewards with power anyone willing to set aside all sense of right and wrong in pursuit of self-advancement, the only people who can stop the insanity are precisely the ones most invested in it. Money buys a lot, including lawyers who tell you the victim is you, never the people you steal from.

As an ideal, capitalism is based on free production and exchange. The reality is *antisocialism,* the systematic encouragement of the otherwise marginal tendency to seek private advantage regardless of externalized cost, rewarding those who steal while punishing those who trust.

The result is a shared mindset or *culture* that propagates craving, suffering, rage and violence. Unlike reason, which must regenerate from scratch with every child, unreason accumulates over generations. Any behavior or belief, no matter how pathological, normalizes over time as the weight of memory lends it legitimacy. Mobs don't act according to empathy; they act like previous mobs.

How do we pivot so readily between thoughtful individuals and fevered fanatics? We might as well ask how an electron can be simultaneously particle and wave. Like grasshopper and locust, the individual and the collective are two sides of the same coin. If mind is a property of individuals, by necessity it's a property of groups. Whether bacterium, insect or human, the individual is only the *individuation* of a shared set of characteristics ranging from anatomical to cognitive.

Like mind itself, madness is fundamentally collective. In contrast to George W. Bush, who fully owned the insanity, Barack Obama doesn't have to be diagnosable to pursue self-destructive policies such as drone strikes that generate more troops in the fight against "globalism," better known as Western imperialism. Given the Roman precedent – and many more like it – maintaining privilege by engaging in permanent war is not likely to end well. Yet our leadership and much of the populace are fully onboard the suicide express.

Though technologically far surpassing ancient empires, the modern West is triggering colossal challenges in ecology and resources. Those who deny the threats have lost touch with the land, water and air that nourish us. "The countries with large and influential indigenous populations," says Noam Chomsky, "are well in the lead in seeking to preserve the planet. The countries that have driven indigenous populations to extinction or extreme marginalization are racing toward destruction."[9]

Who cares if life on Earth is doomed when me and all my pals live in Quantopia?

So hopeless it all seems until we consider that nature freely offers us all the clues we need to find our way back to ourselves. In Peirce's cosmic anarchism, with its spontaneity and freedom from absolute authority, we have our blueprint for a genuine society.

This is a story of evolution, from atom to human to vampire. Should you take this ride, don't be surprised if your polarities flip, your dams pop open and your weapons attack you. The object of this book is to shake up ingrained thinking so as to establish a coherent system of the world ranging from matter to life to consciousness to collective insanity to social revolution.

Strap yourself in.

1. Adams, Mike, "From the front lines of medicine, disillusioned nurse reveals why America's health care system is imploding," *Natural News*, Sept 26, 2013. Available from: http://www.naturalnews.com/042228_health_care_failure_Obamacare_nurse.html (Accessed Dec 31 2013)
2. King, Martin Luther, *Strength to Love*, Cleveland: Collins World, 1977, p 103
3. Hawking, Stephen and Mlodinow, Leonard, *The Grand Design*, New York: Bantam Books, 2010, p 32
4. Crick, Francis, *The Astonishing Hypothesis*, New York: Touchstone, 1995, p 3
5. Hartshorne, Charles and Weiss, Paul (editors) *Collected Papers of Charles Sanders Peirce* (Volume Six) Cambridge: Belknap Press, 1960, p 130
6. Sheldrake, Rupert and Bohm, David, "Morphic Fields and the Implicate Order," *A New Science of Life* (Third Edition) London: Icon Books, 2009, p 303
7. Elsasser, Walter, *Reflections on a Theory of Organisms: Holism in Biology*, Baltimore: Johns Hopkins University Press, 1998

(1987), p 5

8. Smolin, Lee, *Time Reborn: From the Crisis in Physics to the Future of the Universe,* New York: Houghton Mifflin Harcourt, 2013, p 148
9. Chomsky, Noam, "Can Civilization Survive Capitalism?" *CounterCurrents,* March 11, 2013. Available from: http://www.countercurrents.org/chomsky110313.htm (Accessed Dec 31, 2013)

Chapter 1

Physics and Afterphysics

Giordano Bruno discovered in the lights of the night sky a bottomless ocean of suns where others saw only sketches projected from human imagination. Alone among the pioneers of science, Bruno fully absorbed the lesson of Copernicus, something even the solar revolutionary himself failed to grasp. Not only is the cosmos not centered on Earth but the very idea of *center* has no physical meaning. There's no more a privileged location from which all places are subject to objective measurement than a virgin or a goatfish in the sky.

"For there is in the Universe," wrote the itinerant philosopher, "neither center nor circumference, but, if you will, the whole is central, and every point also may be regarded as part of a circumference in respect to some other central point."[1] If Earth seems like the center of all things, that's only because we live on it. To lunar dwellers the Moon is center-stage. It's all perspective.

Bruno never hesitated to announce his relativistic revelation to any and all. For this and other "impieties," the church ordered him burned at the stake on Ash Wednesday 1600.

His successors lacked his penetrating insight. Following Isaac Newton's observation that massive bodies attract each other at a distance, consensus opinion coalesced around the idea of a subtle kind of matter permeating space that mediates the force of gravity much as water mediates waves on the ocean. In the nineteenth century scientists updated this approach with their contention that electromagnetic waves propagate across a "luminiferous aether." Aside from serving as a fixed framework establishing the boundaries and absolute center of the universe, the aether was thought to enable the cosmic machine to operate by contact mechanics, not unlike the contraptions we fashion down here on the terrestrial plane.

By the turn of the twentieth century, the great questions of existence seemed to be dissolving in the magic potion of science. The world had never been so clear, the ground never so solid and dependable.

Since then all center and substance have shattered. Bruno could at least count on God. Now we've got nothing, adrift in a void without reference points. The Great Wall of Certainty has collapsed under its own density. From the other side Bruno confronts us with crackling skin and blazing eye.

Up until 1897 the idea of material substance wasn't generally regarded as a pre-scientific mirage. But in that year JJ Thomson cut the "uncuttable" atom. The solid core of matter turned out to be internally differentiated, with vast empty gulfs punctuated by occasional pinpricks of mass. An electron isn't so much a thing as a field of possibilities across which a "particle" randomly bops around like a speck of static on a TV screen. It's a dance whose steps can be calculated according to a probability wave. Let's say an electron is trapped inside a perfectly sealed container. As it bounces off the walls, its probability wave gradually seeps out to the surrounding area until the electron itself is no longer inside the container.[2]

This is why quantum physicists don't speak of substance. Reality is composed of "information." The randomness of the quantum level averages out to the predictability of the perceptual level. All that is solid melts into stats.

The de-centering of all centers began in 1887 when Albert Michelson and Edward Morley carried out an experiment designed to prove the existence of the aether. Their "interferometer," a box containing a telescope and mirrors set at odd angles, could measure the speed of light on Earth relative to its speed in outer space. Since our planet is in motion, scientists reasoned that the light reaching us from a distant source should be either faster or slower than in the stillness of space, depending on whether we're approaching the starlight or receding. But when Michelson and Morley looked at their results, they found no interference and therefore no difference in the speed of light relative to Earth's motion.

For years their findings puzzled physicists, though Hendrick Lorentz wrote up some interesting equations meant to explain how the aether was somehow still relevant despite the no-show in '87. Not until Einstein came along did anyone see the true weight of the Michelson-Morley results. With a little tweaking of Lorentz's equations, he demonstrated that space has no fixed framework, no center or circumference. As far as the universe is concerned, we are nowhere. Bruno was vindicated.

Whether you're adrift in deep space or breezing along at 185,000 miles per second, light always travels faster, at 186,000 miles per second. light always travels 186,000 miles per second faster. Change your frame of reference and the flow of time changes along with it. Only the speed of light remains constant.

Or so we thought. Light has many different speeds, depending on what kind of medium it's traversing. Water, for instance, slows it down by 75%. In the final days of the twentieth century, researcher Lene Vestergaard Hau imprisoned a beam of light in a frozen cloud of atoms, stopping it dead in its tracks and demonstrating, once and for all, that nothing is sacred.[3]

On the first day God created the heavens and the earth and all the varieties of life. Piece of cake.

On the second day the Creator got a makeover as the Cosmic Mechanic, or at least that was the message brought to Rene Descartes by a certain Angel of Truth during an all-nighter alone with a wood stove.

Things started going downhill on the third day when God was retroactively stripped of all living creation. He could keep the cosmic wrench but had to relinquish all claim to species-creation, an honor now ceded to Mother Nature in the form of evolution by natural selection.

On the fourth day the Almighty lost his title as cosmic creator.

Rather than popping into existence straight out of the mathematical mind of God, the universe began in raging chaos. Whatever the cause of all this atomic and galactic structure, it wasn't a switch pulled by a cosmic engineer at the beginning of time.

On the fifth day God the Lawgiver was summarily removed from office. As physicist Victor J. Stenger puts it, "what we call 'laws of physics' are basically our own descriptions of certain symmetries observed in nature and how these symmetries, in some cases, happen to be broken."[4] *The properties of matter aren't the result of an edict from on high but the random way the original unified force got smashed to bits. Stenger compares it to a pencil standing upright on its tip. Inevitably it falls. If it falls one way, we get one kind of universe. If it falls another way, we get a different set of forces and constants. The pencil just happened to tumble in such a way that suns are forged, hydrogen molecules bond in pairs with oxygen, and people project images onto the night sky.*

On the sixth day the dragon was back from the unholy deep. Long ago, when God was called Marduk, he went around telling everyone he'd banished forever his arch nemesis, Tiamat, to the sea. But Tiamat was just waiting for the right mathematician to come along and give her some mouth-to-mouth (Feigenbaum, 1978). Turns out the simplest little deterministic equation unleashes infinities of self-similar scales of complexity. Which boils down to one thing: we really don't know what's going to happen. The flap of a butterfly's wing could result in a typhoon halfway around the world. Even omniscience regarding all possible atmospheric variables is of no use in telling us whether or not it will rain on Wednesday in Anaheim. God couldn't even make it as a weatherman.

Now comes day seven and the Almighty is tucked away in a rest home, reminiscing on the epic battles of old. The end, we are assured, will come swiftly and peacefully. Only the best for the big guy.

So is that it? No ground of being, no substance or center, nothing true or dependable?

You don't have to consult Aristotle to realize something is

holding all this up. You can't have miles and miles of accident and no essence anywhere in sight. Something's got to be substantial, not just informational. Absolute, not just relative. Even illusion is illusory only in contrast to reality. Who or what is hallucinating this hallucination?

For the answer we must go back, once more, to that magnetic moment when the world turned inside out. The clock is winding down on the nineteenth century as young Henri Bergson, a Polish Jew transplanted to France, studies philosophy at the Ecole Superier Normale. Captivated by the English positivist Herbert Spencer and his book, *Progress: Its Law and Cause,* Bergson is dazzled by the promise of a completely coordinated system of knowledge, a synthetic scheme founded on a single absolute principle: the persistence of force. Physical science, prophecies Spencer, shall render the world transparent, granting unimagined power to the human race.

Then one day Bergson is shaken by a terrible insight, as if the whole twentieth-century intellectual meltdown has appeared to him in a blast.

There's no time in physics.

"Newton's laws of motion," according to physicists Christopher Hill and Leon Lederman, "make no distinction between past and future, and time can apparently flow in any direction." On their website devoted to mathematician Emmy Noether and her principle of symmetry-breaking, Hill and Lederman describe the universe as a movie that could run through a projector in reverse as readily as forward. "When applied to simple systems, billiard balls colliding on the table, atomic collisions, etc., it would not be possible to tell in which direction the film was progressing. The motion we see satisfies laws of motion that are the same, whether run forward or backward."[5] Future and past are effectively interchangeable.

"Notice another peculiar aspect of physics," write Hill and Lederman. "Nowhere in any formulation does the issue of a special point in time called 'now' ever occur. Yet, we humans sense something we call 'now.' Is it an illusion? We call this the 'Now' question."

I can't help but feel present. Even memories concern moments once present. To be human is to be temporal, informed by a past and oriented toward a future. Without ongoing presence our consciousness, the sensation of now, is null and void. Lacking real time, we aren't real either.

In the course of eradicating all substance and foundation, science is getting personal.

Since objects move across space only over time, measurements of motion include temporal as well as spatial readings. By observing time exclusively in terms of events in space, physics demotes time from prime mover to a fourth coordinate on a graph. Since we know the (ideal) speed of light, a given duration can be defined as the amount of space covered by light in that time. A second is thus 299,792,458 meters.

But there's more to time than physics knows. Of all we observe outside us, time is the one thing we also find within. Always a step ahead, always keeping us off balance, time can't be isolated, analyzed, observed from afar. Like fire, it can't be frozen. Nothing can exist without first entering existence. Since "before" and "after" are temporal terms, nothing can precede or follow it.

If time is already complete like a book you could open to any page, if it has no impulse or direction — nothing inside that drives it — how could we? Far from a web of blinking neurons, the substance of the mind is time. To explain ourselves, we must turn to *afterphysics*.

The passage from one ephemeral moment to the next conceals the essence of time: the continual deepening of a singular present. For Bergson, the past is only what's no longer materialized, no longer at the glossy surface of time. The past cannot be seen, but we can evoke its outlines in a process we call *memory*.

Giordano Bruno taught that the universe is boundless and therefore infinite. He was half right. It is unbounded but in the same sense as the surface of the Earth. Since the Earth is round, if you keep going in a particular direction, you'll eventually come back to where you started. The universe is like a balloon with stars and nebulae painted across its surface. As the ball expands, the galaxies disperse.

Reality isn't just the surface of the balloon but the "breath" inside. At the center is the origin of space and matter, the so-called big bang. Between the sensorial surface and the singularity at the core lies the vast deep of cosmic memory.

Only in the universe of time do we find our center. Echoing the ancient philosopher Heraclitus, who posited fire as the primary reality, Bergson found in time the true ground of being, the substance on which all things depend. But the world wasn't ready. By the 1930s the philosophical community, which once idolized him, had largely capitulated to the mechanistic outlook of physicalism. Perhaps it was the sting of abandonment that compelled him to remain with his people and die in Nazi occupied France, a principled death not unlike Bruno's refusal to compromise with the equally deranged Inquisition.

Aside from time itself nothing lasts forever, not even nature's "fundamental constants." Starting in 1928 the speed of light in outer space was consistently measured at 20 kilometers per second less than previous measurements. This odd state of affairs lasted until 1945 when the previous higher value was mysteriously restored.[6] Astrophysicists recently spotted a fluctuation in the alpha constant between 8 and 11 billion years ago, when the light emanating from quasars was slightly dimmer than at any time before or since. It was a laid-back era when electrons were slightly looser in their nuclear orbits and, as a result, emitted radiation at a frequency .001% lower.[7] May not be much but enough to explode eternity forever.

As long as we dream a changeless realm of equations, we can't account for nature's evolving patterns. But what factor, if not eternal law, holds it all in place? Why do molecules and crystals persist in forming the same old ways? Why do suns keep on condensing and aging and detonating, seeding the galaxies with new generations? Why doesn't it all just run off-track?

Not law but habit. Not eternity but memory.

Inspired by Darwin's theory of "descent with modification," philosopher CS Peirce proposed that the basis of order in the universe is not rigid law but a set of habits that evolved into being. Peirce was troubled by the arbitrary nature of physical law. Why this law and not another one? "Law," he wrote, "is par excellence the thing that wants a reason."[8] The struggle to endure provides that reason. As Nietzsche put it, "our mechanical laws would not be eternal, but evolved, and would have survived innumerable alternative mechanical laws."[9] Physicist Lee Smolin refers to this outlook as "cosmological natural selection."[10] By this view transcendent laws, rather than hatched from the mind of Newton's God, are products of our own devising.

Scientists, says theorist Paul Davies, generally view laws of nature as "the bedrock of reality ... the eternal truths upon which the universe is built."[11] By contrast Smolin argues that "just about everything we now think is fundamental will also eventually be understood as approximate and emergent: gravity and the laws of Newton and Einstein that govern it, the laws of quantum mechanics, even space."[12] While space and matter and the laws of physics could very well be emergent, the one thing that cannot emerge over time is of course time itself.

Evolution entails spontaneity, not preplanned design whether theological or mathematical. Disputing fundamental uniformity over time, Peirce speculated that the universe began in chaos, a "confused dream" as he called it.[13] The only law that could bring order was "the tendency of all things to take habits."[14]

The universe did indeed start in extreme agitation devoid of material forms or even radiation. When light finally appeared alongside the atom some 370,000 years after the big bang, its speed was established by chance and has remained roughly the same ever since — not out of eternal necessity but simple habit.

It's all about time. What *happens* is influenced by what *happened*. "For atoms and their parts," writes Peirce, "molecules and groups of molecules ... there is a greater probability of acting as on a former like occasion than otherwise."[15] When supernova-induced fusion belches out atoms of carbon or iron, is the outcome conditioned by timeless law or previous iterations of the same procedure?

The forces and particles now known to physics "resulted from the 'locking in' of new laws," writes Smolin, "when the states corresponding to electrons, quarks, and their relatives first emerged as the universe cooled shortly after the Big Bang."[16] With the first electron came increased likelihood of another. With the second electron, a third was still more likely, and so on, as a groove was laid in the universe of time.

The enduring properties of particles gave them a kind of weight, not in terms of mass but memory. Habit is the operating mode of not just organisms, as Darwin recognized, but nature generally. What applies to species of bacteria applies equally to species of atoms. Peirce's living cosmos is akin to a piano played with the echo pedal pressed down. Notes struck long ago still sound, still informing notes struck now. Prolongation of the past is no barrier to the introduction of novelty.

In contrast to Aristotle, who believed the natural state of an object is rest and that motion results from the action of an external force, Galileo realized that nature is on the go. Barring interference the object carries on as it has up to now. Perhaps inertia applies not only to motion across space but order over time. Any random dust particle retains its motion; only certain kinds of objects retain their form, statically as with atoms and crystals or dynamically as with whirlpools and organisms.

Instead of trying to fit life into the unyielding framework of a cosmic machine, we start with life and work our way back. There's a reason *transpire* means both to happen and to breathe. The totality is alive, the universe its song.

By expelling the cosmic center, Bruno banished us once again from Eden. The world is casually indifferent toward our fate. Now, four centuries later, physics threatens to restore the centrality of human existence in the guise of the infamous *anthropic principle*. Stephen Hawking provides numerous examples of physical properties perfectly tailored for the emergence of Earth and its biosphere. If the "strong force," which binds the atom, differed in strength by only .5%, carbon and therefore protein could not exist. A 4% shift in the strength of electromagnetism would disallow oxygen. Just .2% heavier protons would decay into neutrons, and a 10% change in the "sum of masses" of the quarks that comprise protons would drastically reduce stable atomic nuclei. As Hawking writes, "summed quark masses seem roughly optimized for the existence of the largest number of stable nuclei." Even the number of spatial dimensions is exactly right, as less than three would yield a very boring universe, and four would tweak gravity in such a way that the sun would either fragment or collapse into a black hole.[17]

In response to this dilemma, Hawking espouses the standard line that many universes must exist in addition to ours. If it's a trillion-to-one that our universe is set up just right, most likely a trillion other universes harbor the wrong conditions. Like Goldilocks, we naturally wind up in the one that's not too soft and not too hard.

But then why not appeal to an intelligent designer? The trouble with proposals like God and Goldilocks is that no testable mechanism is offered to back it up. We say a deity did it or chance did it. Maybe the Flying Spaghetti Monster did it. Whatever the pat answer is, that's where discussion ends. The kind of explanation we really need is one that tells us something about nature we didn't already know.

Perhaps the cosmic pencil fell just the right way because this universe isn't the first. Instead, like an organism, a cosmos is a recurring phenomenon expressing a long-established way of ordering things. Every universe that emerged like ours increased the likelihood that subsequent iterations would emerge the same way. Instead of a trillion universes lying side-by-side, the universe was *preceded* by a trillion others, all helping to shape our multi-galactic abode. In contrast to the transcendent model, the habit-evolutionary model *predicts* an anthropic universe. Rather than explain away the problem, we make an actual claim, namely that memory is a property of nature. As we shall see, this is indeed a testable hypothesis.

Over incomprehensible eons the species of universe evolves, and we are privileged to participate in the latest and greatest generation. Deepening grooves increase the likelihood that each organically formed system, from cosmic to atomic, stays on course. Hydrogen keeps pairing up with oxygen because the weight of previous similar events keeps them on track. Snowflakes keep forming their six-fold symmetry because the force of history molds them that way. If snowflakes have no need for an omniscient being or transcendent equations to tell them what to do, neither do we.

The only true God is the one Bruno met, face to face, in the flames.

1. Singer, Dorothea Waley, *Giordano Bruno: His Life and Thought,* New York: Henry Shuman, 1950, p 365
2. Kaku, Michio, *Hyperspace: A Scientific Odyssey Through Parallel Universes, Time Warps and the 10th Dimension,* New York: Anchor Books, 1995, p 117
3. Hau, Lene Vestergaard, "Frozen Light," *Scientific American,* July 2001

4. Stenger, Victor J, "Natural Explanation for the Anthropic Coincidences," *Philo*, Vol. 3, No. 2, Fall-Winter 2000, pp 50-67
5. Hill, Christopher and Lederman, Leon, "Teaching Symmetry in the Introductory Physics Curriculum, VII. Discrete Symmetries in Nature." *Grupo De Fenomenologia de interacciones fundamentales*. Available from: http://gfif.udea.edu.co/emmynoether/discrete.htm (Accessed Dec 31 2013)
6. Sheldrake, Rupert, *Science Set Free: 10 Paths to New Discovery*, New York: Deepak Chopra Books, 2012, p 92
7. Collins, Graham P, "Plus ça Change," *Scientific American*, November 2001
8. Peirce Edition Project (editors), *Writings of Charles S. Peirce: A Chronological Edition*, Bloomington: University of Indiana Press, 2010, p 101 (Vol. 8)
9. Quoted in Sheldrake, 2012, p 98
10. Smolin, Lee, *Time Reborn: From the Crisis in Physics to the Future of the Universe*, New York: Houghton Mifflin Harcourt, 2013, p 125
11. Davies, Paul, *The Mind of God: The Scientific Basis for a Rational World*, New York: Simon and Schuster, 1992, p 73
12. Smolin, p xxx
13. Hartshorne, Charles and Weiss, Paul (editors), *The Collected Papers of Charles Sanders Pierce (Six Volumes)*, Cambridge: Belknap Press, 1960, p 72 (Vol. 1)
14. *Ibid*, p 84 (Vol. 6)
15. *Ibid*, p 223 (Vol. 1)
16. Smolin, p 155
17. Hawking, Stephen and Mlodinow, Leonard, *The Grand Design*, New York: Random House, 2010, pp 159-161

Chapter 2

Absolute Presence

January 8th, 1994, I'm in a crowd of 3000 people watching Kurt Cobain annihilate not only the set of his just completed show but his guitar, in the process tortuously extracting every last bit of sonic data from that final chord. Stepping back for a moment from the general buzz of semi-delirious ecstasy, I wonder why he's doing that. Only many years later do I understand. He's saying to the crowd, look, see that guitar? That guitar is never playing another song. That show you just saw? That was a one time only deal. It was beautiful when it happened, but that moment is irreversibly gone.

Time is all too real. This is why for eons people have tried to find some kind of escape clause. The myth of timeless eternity — where truth never changes and everyone knows your name — is as old as human imagination and as novel as modern physics.

Prior to Einstein, physicists typically regarded "spacetime" as a conceptual convenience. Yet today we are told, on the basis of relativity theory, that time is a static dimension composed of moments set side by side from remote past to distant future, all of them equally present. Think you're progressing through a book? Think again. You're at once on the first and last page and every page between.

As Lee Smolin observes, the idea that we inhabit a four dimensional "block universe" follows from Einstein's principle of the *relativity of simultaneity*.[1] To restore real time we must refute this principle without contradicting the well established phenomenon of time dilation. Rather than abandon relativity theory, we must reformulate it in accord with absolute simultaneity, meaning that a single present moment spans the entire universe.

Einstein's discovery of relativity famously kicked off with a question. What happens to our perception of light when we ride it like a bicycle? There is of course no answer. No matter how fast the fastest rocket soars, radiant waves of electromagnetism still outrun it by the speed of light.

The usual explanation, though not strictly in keeping with Einstein, is that the closer the rocket gets to light-speed, the greater the drag on its passage through time. Slowing in time cancels out the gain derived from speeding across space.

Imagine a group of astronauts who set out to surf the ultimate wave. After chasing light at very high speed, they realize the futility of their quest and reluctantly head back home. But now they have a problem. The drag on their passage through time has left them stranded in a moment prior to the present moment on Earth. Upon their return no one greets them since no one can perceive them. Our senses, after all, monitor the present not the past. Invisible to all except each other, the astronauts are lost not in space but in history.

Obviously this would *not* actually happen. Regressing in time relative to a slower moving object is no more plausible than riding a wave of light. According to not only Einstein but extensive experimental evidence, the travelers would return to Earth in the same present moment as the rest of us, just younger than they would have been had they not embarked on their high speed trip. Instead of reverting to a relative past, the astronauts would find that the duration of their journey has stretched, *dilating* just enough to keep them in sync with the temporal present of Earth.

Why does relativity seem to offer time regression yet deliver time dilation? The phenomenon that needs to be explained isn't so much an effect of relativity as the *absence* of that effect and its replacement with something curiously related but altogether different. A clock that displays an earlier time certainly suggests the act of sliding into the past, but then why not the real thing?

Let's say two rockets are launched just as light is emitted from their point of departure. One rocket travels at 10,000 miles per hour in pursuit of the photons, while the other goes 10,000 miles per second, 3,600 times faster. Because the speed of light *in vacuo* is absolute — always measuring the same regardless of your own speed — the fast rocket recedes from the photons at exactly the same speed as the slow rocket. Something has got to give, and that something is the rate of temporal passage for the fast rocket. The fast rocket stays alongside the other in time but undergoes reduced rate of decay. As revealed by its clock, the fast rocket is "younger" than it ought to be.

Einstein offered no explanation for the substitution of time regression with time dilation. Why does the fast rocket remain in the same present moment as the slow rocket? Why not slip behind?

We know the relativity of simultaneity is wrong because it predicts a nonexistent phenomenon. To get to time dilation, we must restore the universality of the *now*. If all things, regardless of relative motion, are bound by a shared present moment, nothing can revert to a prior moment relative to anything else. Because the high-speed reference-body must remain present to all other bodies, it decays more slowly rather than temporally regressing. Einstein rejected absolute time without considering that time is absolute not in flow, as Newton believed, but in *presence*.

Like the speed of light, the present moment is the same in all frames of reference. By accepting the relativity of temporal flow while recognizing presence as an absolute property of nature, we account for the facts of relativity without eliminating from time the very qualities that make it temporal.

"What we have," as Bergson observes, "is a present which endures."[2] A thing is real to the extent that it stays in the flow, that it remains present. Reality is flowing presence. The question is not whether Einstein was wrong about time but where, precisely, he went off the rails.

Let's say two events occur simultaneously from the point of view of observer A. If A is in steady motion relative to observer B, are the two events also simultaneous to observer B? The answer is no. As we shall see, however, this fact in no way establishes relative simultaneity as an objective phenomenon.

Al and Bob are equipped with superhuman perception. Standing on a railway embankment beside the tracks, Al watches a train rush by. Bob is a passenger on that train. Just as Bob passes Al, lightning strikes the engine car and the caboose. Since the lightning bolts are equidistant from Al, he sees them simultaneously. Since Bob passes Al at exactly that moment, he too is equidistant from the flashes. However, because he's moving toward one flash and away from the other, Bob registers the lightning that strikes the engine before the one that strikes the caboose. As Einstein puts it, "Events which are simultaneous with reference to the embankment are not simultaneous with respect to the train, and vice versa." From this he infers that every frame of reference "has its own particular time."[3]

Einstein's conclusion is unwarranted. His mistake is to treat both reference frames equally in relation to the lightning despite the fact that the embankment is at rest with respect to the flashes while the train is in motion. Because Al, standing on the embankment, occupies the same frame as the lightning, he perceives the timing of the flashes correctly. In motion on the train, Bob occupies a different frame of reference and therefore perceives the timing of the flashes incorrectly.

Einstein's scenario unfortunately muddies the issue by having the lightning strike the train, which might suggest that the lightning and the train share the same frame of reference. This is not the case, however, since the lightning doesn't participate in the train's motion. Nevertheless, to be absolutely clear, let's consider an alternate version of Einstein's scenario. Instead of lightning, imagine a light bulb switches on in the center of Bob's car. In Bob's frame of reference the light reaches the front and back of the car at the same time. In Al's frame, meanwhile, the forward motion of the train causes the light to appear to reach the back of the car prior to the front. Since Bob occupies the same reference frame as the light, he correctly perceives the timing of events. In a different frame, Al misperceives the timing. Rather than claim both observers are equally correct — and therefore occupy different times — we surmise that one is right and the other wrong within the context of a shared present moment.

The appearance of relative simultaneity follows from the fact that light cannot instantaneously transmit from object to eye, whether human or electronic. The relativity of simultaneity is thus a special case of optical illusion. Since the illusion arises from the limited speed of the light rather than its organic processing in the eye, it fools machine as readily as human.

Einstein greatly advanced our understanding by explaining the facts of light without recourse to the luminiferous aether that provides a standard from which the motion of all reference frames can be measured. In the absence of the aether, Einstein concluded that all frames are equally valid. While this is true in the abstract, in practice any given set of events occurs within a given frame, privileging that frame above all others with respect to those events. In Einstein's scenario the frame of the embankment is privileged. If a light appears within the train, the train is the privileged frame in which to measure the timing of events.

A given frame is privileged not in the absolute sense of the fixed aether but only *relative* to a particular set of events. By assuming the equal validity of all frames of reference in every instance, Einstein was insufficiently relativistic in his analysis.

In the absence of a universal frame of reference, no frame is always privileged in relation to others, What Einstein overlooked is that the inverse is equally true: *all frames of reference are always privileged in relation to themselves*. Because the measurement of the timing of events is correct for observers occupying the frame in which the events take place and incorrect for observers occupying different frames, we have no need to resort to the extraordinary claim that different frames occupy different times.

A single present moment encompassing both reference frames is in fact implicit in the train scenarios. In the case of the light flashing at the center of Bob's car, Al says the light reaches the back of the car *at the same moment* Bob says it hasn't reached the front or the back. Later, Al says the light reaches the front of the car *at the same moment* Bob says it has already reached both the front and the back. In order for Al and Bob to disagree on what's happening at any given moment, they must both occupy that moment. Einstein highlighted the discord across reference frames while neglecting the underlying accord.

If an event happens at different times in different frames, different events must be happening at the same time in those frames. To deny this is to assume from the outset that the phrase "at the same time" cannot apply to different frames. Yet this is precisely what Einstein ostensibly proves in his discussion of the relativity of simultaneity. We cannot assume at the outset the very thing we seek to prove.

The fact that time never regresses for one reference frame relative to another is just a roundabout way of saying the past cannot be physically revisited. In his 1994 book, *Black Holes and Time Warps,* theorist Kip Thorne proposes taking a step back in time by adapting a wormhole for use as a portal to the past.[4] If he's right, the present isn't absolute after all. Yet his proposal is clearly wrong, and it's unviable precisely to the extent that he substitutes time dilation with time regression.

In his thought experiment, Thorne acquires a wormhole large enough for a person to pass through. He places the mouth in his living room and the tail in the "family spaceship" parked on the front lawn. His wife Carolee takes the controls while he looks on from the living room. Before she takes off, they reach into the wormhole from opposite ends and take each other's hand.

What's striking about the wormhole is that its internal length is the same no matter how far apart the tail is from the mouth, allowing Thorne and Carolee to continue holding hands throughout her trip.

After launching, Carolee flies for six hours at the cusp of light-speed. She then stops on a dime, turns around and heads home at the same speed before returning to the front lawn. For her the whole trip takes 12 hours. When Thorne notices 12 hours have elapsed, he looks through the wormhole and sees her in the spaceship landing safely on the lawn. But when he looks out the window, there's no spaceship because those 12 hours she spent at high speed have dilated 7300-fold, stretching her half day into ten years for her husband. The tail is now nearly ten years to the past of the mouth.

So he goes on with his life, waiting for Carolee to arrive in his timeline not hers. When the day comes and the spaceship lands, he opens the hatch to greet her. Stepping inside he notices she's holding hands through the wormhole with someone. That someone is of course him. He then crawls through the wormhole and meets himself as he was ten years earlier.

Implicit in Thorne's scenario is that if Carolee exits the spaceship via the hatch, she's in the same present as Earth but younger than she should be by ten years. But if she exits by way of the wormhole, her 12 dilated hours are effectively cashed in for 12 undilated hours, leaving her ten years in the past.

What is it about the wormhole that allows its user to exchange dilated for undilated time? Presumably it has to do with the fact that the mouth and tail of the wormhole are hooked together in time, which Thorne establishes with the hand-holding gimmick. But why would a single timeframe within the wormhole enable this effect? Thorne offers no rationale.

In reality, because the tail dilates in time rather than slowing, it stays in the same present as the mouth while undergoing reduced rate of decay. An age differential is established between mouth and tail but not a time differential. The wormhole carries within it a single timeframe from end to end, but this is also the timeframe outside of it.

Einstein referred to time dilation rather than "time slowing" because without compensatory stretching of the period covered by the high-speed journey, reduced rate of temporal flow would cause the high-speed frame to revert to a relative past, an inconceivable outcome. So let's rerun Thorne's thought experiment with the recognition that time slowing cannot be teased out from time dilation.

Kip and Carolee take each other's hand as she departs on her voyage. But when he looks through the wormhole after 12 hours, she's still on her outbound journey. Realizing the project will not play out as he intended, he withdraws his hand. From her point of view, his hand disappears only seconds into her journey. When she returns, she finds that whether she exits the ship through the hatch or the wormhole, she occupies the same present as everyone else, just ten years younger than she would have been had she not taken the trip.

Thorne's mistake is to replace time dilation with time regression in the context of a wormhole. Einstein's unspoken substitution of relative temporal reversion with reduced rate of decay surfaces in confused form in Thorne's model, which treats the intuitive shadow of time dilation as an actual phenomenon.

Aside from the paradoxes it unleashes, such as pre-empting the time traveler's birth, the possibility of visiting the past nullifies the significance of the conservation of energy, which depends on the orderly flow of time. How can the universe contain the same amount of matter at any given moment if time travelers can subtract the mass-energy of their bodies from the present while doubling it up in the past?

Worse, if a future time traveler could visit our past and make changes with far reaching consequences, everything happening now could retroactively un-happen. To assert the possibility of traveling back in time is to deny the reality of everything we experience, even our own existence. The fundamental truth Einstein overlooked isn't just that time is real but that reality is time.

With the recognition of absolute presence, we account for time dilation without taking on the baggage that comes with time travel. In addition we gain a framework within which to resolve the central mystery of quantum mechanics.

A cat is locked in a steel chamber, its only companion a "diabolical apparatus" consisting of radioactive atoms, a Geiger counter, a hammer and a flask of hydrocyanic acid. A single decaying atom will trip the Geiger counter, releasing the hammer into the flask and fatally poisoning the cat. The probability of an atom decaying is exactly 50%.

Erwin Schrödinger devised this thought experiment to illustrate the fact that nothing definite seems to happen at the quantum level of existence. Unless an observer comes along and peeks inside, Fluffy is neither dead nor alive but in a twilight zone of half and half. "The wave function of the entire system," according to Schrödinger, "would express this by having in it the living and the dead cat mixed or smeared out in equal parts."[5]

Much like trajectory in classical physics, the wave function charts the progress of a quantum system. The difference is that the wave function concerns probabilities not actualities. According to the Schrödinger equation, one wave function morphs into another but never "collapses" into a clear-cut event. Lacking an equation whose solution would collapse the wave function, physicists must resort to brute measurement to find out what's going on down there. So long as no one measures it, a quantum system occupies a *superposition* of multiple states at once, including different locations.

It's one thing for a bunch of invisible particles to occupy contradictory states, quite another for an animal to be neither dead nor alive but in a superposition of both. Whether or not anyone opens the lid to peek inside, the cat presumably occupies one state or the other. As Niels Bohr observed, somewhere between particle and large-scale existence, quantum vagueness cuts out.[6]

To this day the default answer to the quantum dilemma is Hugh Everett's "many worlds" interpretation, according to which every tangible event requires the splitting of existence into a different cosmos corresponding to each quantum possibility. Everett's proposal resembles Einstein's belief that every frame of reference generates a different now. Whereas in my world the lightning bolts are simultaneous, in your world they're not. Everett merely exaggerates the effect: whereas in my world the cat is dead, in your world the cat lives.

The intrinsic nature of time gives us another way of approaching the problem. According to the spacetime model, the future is symmetrical to the past in that both are equally determined in the four dimensional block universe. By contrast real time is asymmetrical. The past actually happened; the future is psychological projection. We can imagine what might happen next, but it has no reality. Instead of passively drawn across the temporal axis of spacetime, we remain always at the leading edge of time, surfing a wave called *now*. But the leading edge of time has some play in it.

As we know from investigations into the process of *decoherence*, not only measurements but any environmental interaction results in a single definite state in place of superposition. Researcher Wojciech Zurek calculates that a particle interacting with its environment has a decoherence time — the time needed to shed its superpositions and resolve into a single location — of 10^{-23} seconds, i.e. a second divided 100 billion trillion times.[7] Unless momentarily plucked from their usual state of superposition, particles occupy an incredibly tiny band of time. The *now* of the electron is not the *now* of sensorial existence.

If a definite state of existence requires the passage of 10^{-23}

seconds, each moment is removed from absolute presence by exactly that duration. As theorist Roger Penrose observes, the quantum is characterized by temporal continuity, which is disturbed whenever a measurement is made, triggering a "jump" to a new set of possible states.[8] In contrast to the continuous development of a quantum system, familiar "classical" existence consists of a sequence of discrete moments, each harboring a definite state of matter.

Within the confines of continually renewing presence, the wave function never collapses, just as the Schrödinger equation stipulates. Measurements or other interactions draw the particle into a complete moment of time, but as soon as the interaction is over, the particle resumes its continuous quantum course. To be tangible and definite, matter requires temporal extension beyond continuous presence. Shifting our perspective from quantum to classical is no different in principle than measuring the properties of hydrogen and oxygen atoms and then, after expanding our field of view, measuring the very different "emergent" properties of water. Tangible matter emerges from the quantum over the course of a cinematic-like succession of moments.

Resolving the quantum dilemma is literally a matter of time. Bohr's divide follows from the fact that quantum existence is absolutely present and therefore continually indefinite. Definitude requires a complete moment of time, which buds off from the continuous present every 10^{-23}, disappearing into the past as the subsequent moment crystallizes.

Whether or not anyone peeks inside the steel chamber, Schrödinger's cat is indeed dead or alive. While the quantum elements comprising both cat and environment can be "smeared" between multiple states at once, the cat as a whole depends on a sequence of moments, occupying a definite state of existence at any given time rather than a spooky combination of opposites in a continuous present.

Quantum mechanics in no way demonstrates that human consciousness somehow co-generates existence. If our species disappeared, the symbolic world of culture and knowledge would of course vanish with it, but Earth and the rest of creation would carry on as usual. Though Einstein wanted a reliably deterministic physics, most fundamentally he sought realism. Insisting that physics describe a world indifferent to our perception of it, his chief problem with quantum theory was that the act of determining an event merely by observing it seemed to abolish independent objective existence. If, however, superposition is a function of absolute presence, realism is restored. It *really is* indefinite down there.

From the standpoint of relativity, presence is absolute only in the spatial sense of applying to all frames of reference. But spacetime cannot capture real time. At the quantum level, time is revealed as a thing in itself apart from space, its defining trait flowing presence.

There's no time like the present. From this singular insight follows a view of nature and ourselves at once novel and perennial.

1. Smolin, Lee, *Time Reborn: From the Crisis in Physics to the Future of the Universe,* New York: Houghton Mifflin Harcourt, 2013, p 55
2. Bergson, Henri, *The Creative Mind,* New York: Philosophical Library, 1946, p 180
3. Einstein, Albert, *Relativity: The Special and the General Theory,* New York: Crown Publishers, 1961, p 26
4. Thorne, Kip, *Black Holes and Time Warps,* New York: W. W. Norton & Company, 1994, pp 500-504
5. Whitaker, Andrew, *Einstein, Bohr and the Quantum Dilemma: From Quantum Theory to Quantum Information,* Cambridge: Cambridge University Press, 2006, p 237
6. *Ibid,* p 413

7. *Ibid*, p 310
8. Penrose, Roger: *The Emperor's New Mind*, Oxford: Oxford University Press, 1989, pp 227, 250-251

Chapter 3

Merry Go Round

At the Infinite Emporium all the consumer items you've ever craved are on display down aisles stretching beyond a fluorescent horizon. The merchandise is infinite not only in quantity but intelligence. Every TV already knows what you want to watch. Tucked away inside every cellphone are all the friends you'll ever dial up in desperation at three in the morning. Your fabulous career awaits in the higher education aisle. A spouse in a house is yours on layaway. Your future children are boxed up in the back, whole lifetimes ready to be scanned and bagged and carted off to the bottomless trunk of a waiting car. It's a wondrous store to be sure. But once you buy in, your every thought and feeling comes with a barcode.

Quantopia is a prison, a *closed system* in the lingo of physics. Under the benign rule of deterministic equations, the course of every process is set, the entire sensory realm a mere shadow of timeless mathematical dicta. Just as shadows have no say over their movements, we the people are helpless to give direction or meaning to our lives.

"Everything in our world," says cosmologist Max Tegmark, "is purely mathematical — including you."[1]

The centuries-long retreat from openness and freedom to suffocating determinism comes to fruition with the spatialization of time. In the block universe, temporality is extinguished. With time laid out end-to-end like the familiar dimensions of space, the future has no more play in it than the past. Instead of perched at the leading edge of time — ready to set our own course — we're dragged helplessly across a prefabricated timescape, our unshakable sense of time's flow written off as subjective illusion.

From the transcendent Sphere posited by Parmenides to the

"amplituhedron" of contemporary physics, we are told that everything in creation boils down to a perfect and immutable geometrical object. And while the model of the amplituhedron can indeed predict the outcome of any and all particle interactions, it can't answer the simplest question of them all: why this now?

The alternative to static idealism is evolution, not just biological but physical. Instead of being ruled from above by disembodied principles, material forms emerge spontaneously from the ground up. Instead of determining the ways of nature, mathematical laws simply echo tendencies ingrained naturally over time. Cosmic order stems from habits forged by a living universe that's making it all up as it goes along.

It's a battle for the ages. In one corner is Plato, goliath of geometry and idealism, ready to land a blow for timeless eternity. In the other corner stands a scrawny but resourceful Darwin, defender of creatively evolving nature.

Bergson's stroke of genius was to rescue the absolute from its timeless domain and restore it to the flow of actuality. To be absolute is to exist intrinsically, defined not by others but in relation to oneself. Time needs no justification or cause outside itself. So too a frog exists intrinsically, in and for itself. A computer, on the other hand, is a structured assemblage of atoms capable of carrying out programmed tasks for human users. It has a physical inside but no interiority, no self-reference. Unlike the frog, the computer doesn't care if you boot it up or pull the plug.

The now of right *now* is the same now it's always been, though with each hour or day or year that passes we call it something different. In the context of forever now, the underlying forms of this ever shifting material world can't help but pile up in a kind of cosmic memory bank. Time isn't passing so much as accumulating.

Absolute presence is the generalization of all the events that ever peeled off the main body of time. Like a tree dropping leaves into a rushing current, the intrinsic present remains standing even as it particularizes into fleeting moments. Unlike the "memory" found in fossils, whether biological or astronomical, the natural memory intrinsic to absolute presence isn't so much a relic of the past as the past itself folded up, hidden in the present.

This, in essence, is the theory of nature proposed by esteemed twentieth-century quantum theorist David Bohm. Concerned that we seem to be caught between two realities, the material and the mental, Bohm sought to bring both poles of existence within a single conceptual framework, an "unbroken totality of movement" capturing everything from photons to the light of consciousness.[2]

His solution involved a kind of merry go round. Implicit in what we observe is a deeper order of existence, silent and whole, which is continually translated into the sensuous particular and back again.

It's a common theme of life. To get rain we need covert evaporation, and to get evaporation we must have overt rain and so on.

Articulating a thought requires breaking it down to a linear sequence of words. A thought is a totality that exists all at once, but we don't know how to communicate it in that form, so we tear it apart into words the listener must convert back to a thought. To reach a third person, once again the crystalline perfection of the self-contained idea shatters into words so as to reconstitute in the mind of the next listener. And round and round it goes.

Like a wave on the sea, the world of matter and space is projected from a covert "implicate" order. And like water dropping back into the oceanic whole as the wave passes, the outer "explicate" order is re-injected into the implicate. This back-and-forth action, says Bohm, is the beating heart of the cosmos.[3]

What goes out returns. Underlying order unfolds into sensory existence, which instantly folds back to its core nature. A seed opens into a flower that generates more seed before wilting and returning to the soil, nourishing its progeny. To be audible on a radio, a song must be folded into a signal that modulates a radio wave and then unfolded back to sound waves. As tangible phenomena, evaporation and seeds and radio waves are explicate but nonetheless analogous to the action of the implicate order.

In the process of cycling through the implicate, a piece of outer existence is left behind for posterity. "There seems to be at least a tendency," says Bohm, "not necessarily an exact causal relationship, for a certain content in the past to be followed by a related content in the future."[4] Contrary to the physicalist assumption that nature's elementary structures are preserved by timeless principles, Bohm suggests the properties of particles and atoms persist because their intrinsic forms are "fixed in cosmic memory."[5]

Decades before Bohm presented his theory, CS Peirce observed that physics provides no explanation for how the elements remain stable over time. "As for the distribution of potential about an atom or corpuscle, it is regular; but there is no ulterior reason for that regularity, or, at least, none is probably discoverable."[6] Why posit a transcendent law to keep atoms in line when simple habit suffices?

The fundamental forms of nature, according to both Peirce and Bohm, evolved from scratch without the benefit of pre-existent design. As Peirce put it, the "pure arbitrariness" of events in the remote past "would have started the germ of a generalizing tendency … and from this, with the other principles of evolution, all the regularities of the universe would be evolved."[7] In Bohm's terms, any random forms that materialized from primordial chaos immediately folded into the implicate order. Eventually a kind of negative image of each type of particle and atom built up, ensuring the endurance of those forms in the sensory world.

More precisely, what accretes in the implicate order isn't so much a picture of each kind of organically formed object but an ensemble of its possible states. Atoms display different properties at different times because only certain properties are unfolded at any given moment.[8] Bohm compares it to a piece of music, the entirety of which is enfolded in the composition, though at any time only a single set of sounds is playing.[9]

Long after the emergence of hydrogen and helium in the very early universe, larger atoms were forged inside exploding suns. Each novel type of atom would have turned up by chance, but once a set of properties was established in the implicate order, all matter subject to the same set of solar forces would most likely have emerged the same way out of habit. Like the frog, the atom has interiority. It wants to be structured a certain way.

Bohm's model entails two forms of causality. For purely explicate phenomena such as water running down a creek, one event leads directly to the next. Causation becomes complex when the implicate intervenes, that is, when a given process is informed by past similar processes. Rather than dictated from beyond time, the formation of carbon atoms during a supernova follows the pattern established over time in earlier supernovae. Whatever is most relevant from the remote past remains present to all organically enduring phenomena, whether species of particle or species of life. As the originator of semiotics, Peirce might have interpreted the implicate order as the source of signs that instruct natural processes how to proceed, much like street signs.

Niels Bohr identified Peirce's "distribution of potential about an atom" as the orbital pathways of electrons. The number of possible orbits is fixed, regardless of how many are occupied by actual electrons. Whereas increased energy actualizes orbits, energy loss causes electrons to drop closer to the nucleus, allowing higher orbital pathways to revert to virtual status. Yet these orbits remain as real as the occupied orbits, since the current state of the atom is only an explication of the ensemble of possible states in the implicate order.

Like Einstein before him, Bohm was well aware that no fundamental substance props up matter. Since the discovery that the atom is divisible, nothing tangible has come along to replace it as the basis of matter. Instead we're confronted with scores of ephemeral particles that routinely transform into one another or get annihilated in the course of collisions. Not exactly a ground of being.[10]

Einstein explained matter in terms of energized fields that occupy all space. Mostly these fields are spread thin, but wherever energy is sufficiently concentrated we get mass. Though a leading protégé of Einstein, Bohm took a different tack. Traditional physics, he observed, treats the spatial world as the entirety of reality, thereby limiting itself to things that exist outside of each other and interact across space. The meaning of the explicate order is *outsideness*. To complete our picture of reality we need *insideness*. Though matter is spatial and localized, its basis is the implicate order that acts everywhere but can't be pinned down to any particular locale.[11]

As researchers have demonstrated in numerous laboratory experiments, particles in a quantum system exhibit *nonlocal* connection regardless of their distance from each other. This is especially clear in the case of particles with a common history. When a high-energy photon is fired into a crystal that converts it into a pair of low-energy photons, the descendents remain correlated or *entangled*. Tweak one photon, such as its momentum or polarization, and its sibling is altered in a complementary way.[12]

Entanglement operates in complete disregard of the light-speed barrier. Since transmitting the status of a photon to its sibling across space would take time — and the effect is instantaneous — the status of the first photon is somehow already contained in the second. This makes no sense unless we consider that each photon is merely a projection of the implicate photon. The second photon knows what the first is doing because, as individuations of the same totality, they are *in essence* one and the same. To regard photons as separate entities is akin to regarding waves on the ocean as independent.

Same goes for superconductivity. Under the right conditions, electrons "go round obstacles 'cooperatively' without being scattered or diffused," which seems impossible until we consider that "each electron acts as a projection."[13]

Bohm's model gains further support from Richard Feynman's interpretation of the famous "double slit" experiment, in which electrons are fired at a screen which is blocked by a barrier with two tiny openings. If one of the slits is closed, the electrons travel through the one that's open. We might surmise that when both slits are open, half the particles will travel through one slit while the other half go through the other, but that's not what happens. Instead the screen registers an interference pattern, much like the pattern produced by waves on a body of water.

Feynman's approach is called "sum over histories." Since quantum particles typically occupy a state of superposition, Feynman hypothesized that a given electron isn't limited to a single path from point A to point B but instead takes every possible path. One path might be direct, such as going through a slit and hitting the screen, while other paths involve elaborate detours, like passing through one slit and returning through the other only to take the first one again. When the electron finally hits the screen, all its possible histories exist in superposition.

By adding up every path of the electron, physicists arrive at exactly the interference pattern revealed on the screen. As Stephen Hawking points out, Feynman's approach has proved more useful in fundamental physics than the particle-wave model and thus seems to take us deeper into nature's secrets.[14] The apparent jumble of waves is the conveyance of the electron's multifaceted history to the present. At the moment of contact with the screen, the electron's past is made visible. Bohm's model explains this outcome according to retention of the past in the implicate order. As with any intrinsically structured object, the electron's past is *implicit* in its present.

Whether entanglement, superconductivity or sum over histories, Bohm's model fits the profile of quantum mechanics better than any other. But the real kicker is how it bridges the gap between the inanimate and the living. What good is a theory of existence if it can't encompass the theorist?

Like an atom and its orbital pathways, an embryo is as much spectral as tangible. Only in the course of development does it fill in its virtual elements. An adult organism is akin to an atom energized to the point where all its orbits are actualized. Yet even the adult retains a trace of the virtual, otherwise known as *mentality*. Your mind is your own piece of the implicate pie. Your decision to stand up explicates as the muscle contractions that get you to your feet. Reversing the commonplace view, William James famously claimed we're sad because we cry. But this is only the movement from explicate to implicate. The merry go round also turns emotion into tears.

A loyal Einsteinian, Bohm associated time with space and therefore relegated it to the explicate order. However, he defined time solely in terms of moment-to-moment passage, neglecting its flipside, the *now* that neither comes nor goes. Fleeting explications mask a deeper order of continuity, an enduring present repeatedly instantiating as tangible events.

Bohm encapsulated the back-and-forth action of folding and unfolding in the term *holomovement*. But we don't need a neologism here. Unfolding from implicate to explicate is simply the flow of time, while folding back to the implicate is the accretion of memory. Despite fleeting materialization, the form itself sticks around and influences subsequent explications. Where Bohm says "the whole implicate order is present at any moment," he might as well say the presence of the implicate order is absolute.[15]

Quantum existence straddles the boundary between continuous presence and discrete moments, in other words between implicate and explicate. Ordinarily subsumed to the implicate – where all possibilities exist in superposition – a quantum system will settle on a single explicate state when the local environment intervenes, as in the case of a measurement. Environmental interaction is like a bubble forming around a quantum system and carting it off to a realm of definitude. When the interaction is over, the bubble pops, releasing its quantum contents back to the implicate sea.

Explicate time is a series of complete moments bearing definite states of matter. At its core, however, time is always incomplete. If you could eat it like an avocado, instead of the seed in the center you'd arrive at a void, a pulsating nothingness out of which new fruit continually emerges. Try as you might, you can't quite consume that void. Before long you realize it's eating you.

In space everything is given. Take a look around and you'll see four walls adjoining a floor and a ceiling in a tidy arrangement that keeps you cozy and warm. In the context of time, however, one of the walls has a gaping hole, and it's through this opening that the future continually blows in. Part of the world is missing.

The meaning of time is that existence is never given once and for all but always leaves room for novelty. The void at the heart of the world is what brings it to life. Time is the great animator. As post-Bergsonian theorist Gilles Deleuze puts it, "if the whole is not to be given, it is because it is the Open and because its nature is to change constantly or to give rise to something new, in short, to endure."[16]

The universe is fundamentally undetermined because the window is always Open. Confusing reality with a fanciful realm closed off from duration, TH Huxley claimed that if scientists had been around to observe the primordial "cosmic vapor" and measure "the properties of the molecules of that vapor," they could have calculated "the state of the Fauna of Great Britain in 1869."[17]

We must forgive determinists their blind spot. After all, the open is precisely that piece of the jigsaw puzzle that can never be found.

Aside from the ephemeral "now" of the senses — already replaced by the next "now" — we have the true now of absolute presence, the continual wrenching open of existence. Space and matter follow absolute presence like the tail of a comet, always a little off, always a sliver to the past. If you reach out, you can almost touch it, but it's not quite there because it's not in space but *prior* to it. The entirety of physical existence, everything you've ever seen or touched, is but a wake.

Like the breath before speech, absolute presence makes room for spontaneity. What Bohm adds is the other half of the circuit. Time is not only the opening up of existence to possibility but the endurance of an abstracted past, which serves to guide our moment-to-moment materializing present.

Behind the fleeting sensorial moment lies the absolute, open and virtual, just out of reach yet always surfacing with a message from our past.

As the atom is informed in the ways of molecular couplings, the human is informed in the ways of social configurations, the only difference being the degree of freedom to set aside tradition in favor of possibility. Absolute presence is the foundation for change against the ever-humming backdrop of stability — in a word, *evolution*.

As a resident many years ago at Oxford, Bohm swapped insights with a young mathematician named Roger Penrose, now among the world's leading thinkers. Like Bohm, Penrose wants to bridge the divide between the physical and the mental, and he finds a clue in the emerging crystal. The natural state of a "crystalline configuration," as Penrose observes, is whichever setup harbors the smallest amount of potential energy. Just as a rock would rather be on the ground than ten feet up in the air, a crystal is most content when it fits together with the least amount of potential tightening up. Growing by one atom at a time would lead a quasicrystal to a dead end of excess potential energy.[18]

Yet this doesn't happen. Instead the emerging crystal follows exactly the right path to the minimum energy state. How does it "know" which way to go?

With a nod to the quantum, Penrose suggests that different versions of a potential crystal coexist at once. Rather than adding each atom individually to a growing lattice, numerous virtual versions of the completed crystal stack up in superposition, allowing the most energetically correct configuration to be selected from the get-go. No need for second drafts.

Penrose wonders if the same principle is at work in the brain. In support of this conjecture, he quotes Mozart on the experience of composing music. "It does not come to me successively... but in its entirety." Rather than a steady accumulation of notes, the composer "hears" the melody as a whole and only later fills in the details.[19] Perhaps crystallization, whether of minerals or insights, is quantum-driven.

Recent research lends support to Penrose's insight. Superposition has now been implicated in photosynthesis, the process of harvesting sunlight for energy. Since energy dissipates on the way to the "reaction center," where it's captured in chemical form for later use, finding the fastest pathway is paramount. With numerous paths stacked up in superposition, the optimum route can be selected in advance. Yet this should be impossible in the warm buzz of the living cell, which provides way too many opportunities for environmental interaction and the resulting decoherence. Protein vibrations, for instance, which ought to interfere with superposition, instead help preserve it.[20]

How can quantum coherence, which includes both entanglement and superposition, persevere in the organism? Perhaps the answer is that life itself is an evolved form of coherence. Like a tree raising nitrogen and minerals from the earth to circulate in its branches, the implicate order conveys the natural propensity for coherence from quantum ground time to the messy and expansive realm of lived time.

Just as the virtual electron is both a product of actual electrons and a force for conserving their properties, the virtual sunflower both echoes actual sunflowers and serves as a guide for newly germinating seed. Environmental interaction destroys coherence only if the environment is itself incoherent. As an evolved expression of the implicate order, the organism has the same coherence of any quantum system.

How else to explain the striking resemblance of imagination to superposition? Is it just coincidence that nature provides the perfect precedent for our capacity to entertain numerous possible responses to a given situation? The ability to weigh different options before taking action has long been assumed to be rooted in the structure of the nervous system, but those roots may extend far deeper than we ever suspected. When animals go into superposition, it's called *thought*.

Consciousness is nothing if not flowing presence. You might say each of us carries a piece of the absolute in our back pocket. We've come a long way from the electron and the atom yet remained firmly in the fold every step of the evolutionary way. If connectedness to the absolute yields the capacities of wonder and memory, i.e. an open mind and virtual retention of the past, why not imagination as an evolved form of superposition? Why not love, for that matter, as an evolved form of entanglement?

As material forms evolve, so do the virtual forms associated with them. Over the course of evolution the key development has been the weakening of the necessity to follow our virtual guide. As Bergson says, the human brain is a "reservoir of indetermination."[21] Between sensory and motor neurons lies a vast field of synaptic play in which to formulate alternatives to habit. Our capacity for self-determination, though manifest neurologically, is ultimately rooted in the openness and creativity of the universe at large.

Though Bohm's equal in creative insight, Penrose is wedded to the Quantopian myth of mathematics as a conduit to a transcendent realm of law that determines activities on our lowly terrestrial plane.[22] Equations are "timeless entities" and mathematicians "merely explorers in a world that lies far beyond themselves."[23]

At first glance math seems to be an elaborate solipsism. Two plus two add up to four because each term in the equation is defined in such a way that the outcome is a given. Yet somehow, in the course of playing with numbers and operators, the mathematician stumbles onto something profound and universal. Like a songwriter who hits upon a tune that seems like it's been around forever, mathematical investigation has an almost magical power to tap into the depths of nature. Researchers didn't just stumble onto quantum entanglement, for instance, but went looking for it after it was predicted by the mathematics. Examples such as this abound, and it's tempting to concede the point to Penrose.

But the model of cosmic evolution opens up another option. We can think our way into the marrow of nature because we're products of the same evolutionary process that gave the universe its myriad forms to begin with. What we call reason is simply the action of nature applied to human consciousness. "It is as though nature herself," writes Penrose in *The Road to Reality*, "is guided by the same kind of criteria of consistency and elegance as those that guide human mathematical thought."[24] For Penrose this means nature's order follows from a Platonic realm of mathematics that humans have learned to tap into. But it could also mean we developed math simply by thinking naturally.

A single thread of evolution leads, unbroken, from the original cosmic plasma to conscious speculation on the nature of existence. In the sensory realm everything seems disjointed, but an evolutionary pathway is implicit. All intrinsic forms are connected in a treelike structure from particles, atoms, organic compounds and cycling chemical systems to bacteria, eukaryotes, animals, vertebrates, mammals, primates and people. At every step of the way is the impulse of time, this ongoing presence that animates one form and another and still another until it's human thought retracing our cosmic lineage. Far from a complex property of the brain, consciousness is the living presence expressed through it. Via the absolute we are plugged into all the organic forms preceding us.

The role of mathematics, by this view, isn't to reveal an independent realm of quantitative objectivity but to awaken our memory of the patterns of activity upon which our own organic form is founded. Math is a way of stirring up memories that go deeper than our species, deeper than life, perhaps even deeper than the universe.

Time is of the essence. In a fundamentally temporal reality, atoms and forces are creatures of habit. What we call life is only a streamlined sped-up version of physical evolution, and the further life develops, the greater the tendency to generate new tendencies. This process is still at work in the evolution of human cultures, where systems of belief compete in conceptual environments. The notion of a truth beyond time, for instance, is very attractive and for that reason has dominated human thought for eons.

The implicate core is the collective basis of explicate individuation. Like atoms of a given kind, we too are fundamentally *one*, and our underlying singularity was evident in the common culture forged by the pioneers of our species. Like the veins of a leaf, development of distinct cultures and languages over scores of millennia extended the tree of living presence into unforeseeable patterns of differentiation.

Part of the power and resilience of the Platonic mindset follows from its close approximation of reality. Organic form does indeed exist apart from matter in a virtual state. The only mistake is to remove time from the picture. Instead of evolutionary abstraction, the virtual is plucked from the flow and set on a pedestal.

A world under the control of eternal edicts ought to be inert. Instead we're confronted with change and rhythm. Rainfall and evaporation, exhalation and inhalation, aging and growth, all these cycles echo the primordial intercourse of explicate and implicate, object and subject, material and virtual. Every living form is built atop the cyclical activities of simpler forms, every species a new tune and every organism a variation on that tune.

The world came into being of its own accord, a product of spontaneous play coupled with universal habit. No transcendent principles wait in the wings to stamp their patterns onto passive matter. All that's needed is an opening, and nature springs to life.

1. Tegmark, Max, "Mathematical Cosmos: Reality by Numbers," *New Scientist*, September 14, 2007, pp 38-41
2. Bohm, David, *Wholeness and the Implicate Order*, London: Routledge & Kegan Paul, 1980, p 172
3. Sheldrake, Rupert and Bohm, David, "Morphic Fields and the Implicate Order," *A New Science of Life* (Third Edition) London: Icon Books, 2009, p 302
4. *Ibid*, p 303
5. *Ibid*, p 305
6. Hartshorne, Charles and Weiss, Paul (editors) *Collected Papers of Charles Sanders Peirce* (Volume Six) Cambridge: Belknap Press, 1960, p 82
7. *Ibid*, p 26
8. Bohm, p 183
9. *Ibid*, p 200

10. *Ibid*, p 49
11. *Ibid*, p 173
12. Whitaker, Andrew, *Einstein, Bohr and the Quantum Dilemma* (Second Edition) Cambridge: Cambridge University Press, 2006, pp 263-272, 281
13. Bohm, p 189
14. Hawking, Stephen and Mlodinow, Leonard, *The Grand Design*, New York: Bantam Books, 2010, pp 75-76
15. Bohm, p 154
16. Pearson, Keith Ansell, "Duration and Evolution: Bergson contra Dennett and Bachelard," from *Time and the Instant*, edited by Robin Durie, Manchester: Clinamen Press, 2000, p 169
17. *Ibid*, p 153
18. Penrose, Roger, *The Emperor's New Mind*, Oxford: Oxford University Press, 1989, pp 436-437
19. *Ibid*, p 423
20. Cofield, Calla, "Proteins boost quantum coherence in bacteria," *physicsworld.com*, January 11, 2013. Available from: http://physicsworld.com/cws/article/news/2013/jan/11/proteins-boost-quantum-coherence-in-bacteria (Accessed Jan 1 2014)
21. Pearson, Keith Ansell, *Philosophy and the Adventure of the Virtual*, London: Routledge, 2002, p 77
22. Penrose, *The Road to Reality*, New York: Alfred A Knopf, 2005, p 18
23. *Ibid*, pp 13, 17
24. *Ibid*, p 60

Chapter 4

Analysis of Russell

"It often turns out important to the progress of science," writes Bertrand Russell, "to remember hypotheses which have previously seemed improbable."[1]

If only he'd been true to his word. On the brink of a genuinely scientific account of the mind, he cobbled together a straw-man substitute and promptly set it alight. His rejection of "mnemic causation," the influence of the deep past over the present, was intended to clear the way to a materialist concept of mind.

A series of lectures published in 1921, Russell's *Analysis of Mind* is geared around the proposal that the mind has no existence apart from sense data. "All psychic phenomena are built up out of sensations and images alone," he writes.[2] "Beliefs, desires, volitions, and so on" turn out to be "sensations and images variously interrelated."[3] Images may seem more mental than tangible, but according to Russell they "have a causal connection with physical objects, through the fact that they are copies of past sensations."[4] Images reduce to sensations, which in turn reduce to the meeting of nerve endings with the external world. From mind to matter in a few easy steps.

Recognizing that modern physics renders the concept of matter as mysterious as mind, Russell asserts that both terms ultimately reduce to a deeper "neutral" substance. Given his caution as a philosopher, it's no surprise he never completely forecloses on the possibility of mnemic causation.[5] Despite the window dressing, however, *Analysis of Mind* amounts to an attack on the idea that mentality is intrinsically real.

Russell denies that an animal's search for food can be ascribed to its "mental state, which we cannot observe," arguing that its apparent hunger is only an "observable trait in the bodily behavior ... not some possibly mythical and certainly unknowable ingredient of the animal's mind."[6] To say animals want to eat is akin to saying "rivers 'desire' the sea." As in the case of water flowing downhill, "if we knew more about animals, we might equally cease to attribute desire to them, since we might find physical and chemical reactions sufficient to account for their behavior."[7] People fare no better in his analysis. "We may regard a human being as an instrument, which makes various responses to various stimuli."[8] Will, he says, is a mirage generated by the "kinesthetic sensations" that accompany muscular movements.[9]

With self-existence reduced to mechanics, psychology differs from physics only insofar as physics deals with "a given object from different places," while psychology concerns "different objects from a given place."[10] That place is of course the brain, the physical location of subjectivity, a phenomenon equally evident in "the photographic plate."[11] Russell might have done well to take another look at psychology, as anyone who reduces emotion to a "confused perception" clearly has some inner work to do.[12]

The chief threat to Russell's reduction of mind to matter came from his arch rival, Henri Bergson. In his 1911 book, *Matter and Memory*, Bergson asks why, if images are faded copies of prior sensations, we never confuse the recollection of a loud noise with the sensation of a soft one.[13] Unable to answer Bergson's question, Russell can only observe that we have a "belief-feeling" that a remembered image relates to the past.[14] On what basis do we arrive at this belief-feeling? Russell cannot say. How do we acquire our sense of *pastness*?

The job of the brain, according to Bergson, is to calculate possible actions in response to sensory data.[15] Inputs are converted in the most efficient possible way to outputs. That's all there is to it. Within those cerebral folds, you will find no representations of the world, no emotions, no thoughts, no desires, no psyche. For Bergson, locating the qualities of mind in the brain amounts to a kind of neural mysticism. Is the brain so special as to simultaneously participate in the physical world and yet step outside it to represent it?[16]

Rather than construct images of the world, says Bergson, our brains merely facilitate our perception of it. Because the brain does its job, we apprehend (roughly) what is around us. Just as we see the world itself rather than a neural reconstruction of it, Bergson argues that in memory we perceive the past, if only in outline. But how can we perceive something that's no longer there?

"The past has not ceased to exist; it has only ceased to be useful."[17] Bergsonian time is unbroken *duration* that conveys into the present all that preceded it. "Our most distant past adheres to our present and constitutes with it a single and identical uninterrupted change."[18]

The continuous time of the quantum, as expressed in Schrödinger's wave function, is disrupted via interactions with the local environment. From this we surmise that large-scale existence lies beyond the continuity of the enduring present. As a macroscopic object, the brain is indeed limited to the current moment. By contrast, the mind reflects time as it is, in which past (memory) adheres to presence (consciousness). Because the mind is absolute presence filtered through eons of physical and biological evolution, we possess the power of memory, to take an event no longer materialized and re-present it.

Unable to pinpoint where Bergson's proposal went wrong, Russell conjured mnemic causation, not quite what Bergson actually said but close enough that in refuting it, he would seem to have shaken off his nemesis without even mentioning him by name. Alluding to the work of German zoologist Richard Semon, Russell explains his idea. "Whenever the effect resulting from a stimulus to an organism differs according to the past history of the organism, without our being able actually to detect any relevant difference in its present structure, we will speak of 'mnemic causation.'"[19]

A child who has been burned, says Russell, reacts differently to fire than a child with no such experience. If the memory of being burned leaves no trace in the brain, but the child nonetheless reacts to fire in accord with prior experience, this indicates the direct influence of the remote past over the present with no material intermediary.[20]

By proposing that mnemic causation is indicated by the absence of any neural change reflecting a prior event, Russell rigged the results in advance. As we now know, and as Russell surely anticipated, the brain harbors "memory traces" correlated with past events. By materialist assumption, these neural configurations record the past. It may not work exactly like magnetic tape, but the result is the same.

A logician by training, Russell should have realized that mnemic influence in no way implies the absence of "any relevant difference" in brain structure. This is the inverse of the fact that the brain's necessity for the act of recall falls short of sufficiency. Russell makes this point himself, observing that our dependence on brains for memory doesn't prove that recollection is a strictly neural process or that memories are stored in brain tissue.[21] So too the action of the distant past on the present, even if necessary to account for memory, still leaves a role for the brain.

Russell's plan seems to have been to dispose of Bergson's past-within-a-present so as to arrive at Semon's concept of the *engram* as the only possible explanation of memory. A kind of neural engraving, the engram is the change in the brain's resting state following an event such as being burned. It's the engram that makes the child more alert and therefore less likely to be burned again.[22] Semon's explicit denial that engrams could be regarded as "immaterial or metaphysical" must have been music to Russell's positivistic ears.[23]

By attributing "mnemic phenomena" exclusively to the engram, Russell could fully incorporate memory into neurophysiology. Like knowledge, images and habits, memories exist only when aroused from the brain by the appropriate stimulus.[24] As opposed to a mind obeying the laws of mnemic causation, we have a brain governed by "causation of the ordinary physical sort."[25]

Semon, as it happens, wrote the book on mnemic phenomena (taking his cue from Mnemosyne, goddess of memory and mother of the muses). As he writes in *The Mneme*, "Already existing engrams are never remolded but remain as they were first imprinted."[26] The engram's defining trait, stability over time, not only accounts for memory but helps explain the general stability of the organism in the face of the dominant tendencies of transformation and evolution.[27]

Yet it's precisely their stability that makes engrams wholly unlike anything neuroscientists have actually uncovered. Every time someone remembers an event, the relevant memory trace loses its structure and must be "reconsolidated" from scratch. As John McCrone explains in *New Scientist*, "Resurrecting a memory trace appears to render it completely fluid, as pliable and unstable as the moment it was first formed, and in need of fixing once again into the brain's circuitry."[28] If something interferes with reconsolidation, such as high voltage current or a protein-blocking drug, the memory can never be accessed again. This finding, which has been consistently replicated, baffles researchers since it means a memory, once recalled, is lost to the brain and must be re-established on the basis of nothing more than the actual recall, however cloudy, of the past event itself. Just when we become conscious of it, the memory is irreducible to information encoded in the brain. How can this be?

Regardless of how hard scientists try to impress memory into gray matter, it pops back up, bobbing on the ethereal sea of mind. Though facilitating recollection, the brain does so without storing and retrieving information about the past. And why should it? After all, the whole point of remembering something is that you don't have to look it up. To construe the brain as an organic reference library is to banish memory and replace it with mere information storage. The fact that recollection may indeed revitalize past perception is only a problem for the materialist outlook.

Semon proposed another concept, known as *mnemic homophony*, that accounts for memory far better than the engram. Russell praises Semon for this ingenious idea, not for its application to habits and recollections but its explanation of how the richness of experience is rendered into abstractions, a process that befuddled philosophers as diverse as Berkeley and Hume.[29]

Semon compares the emergence of abstraction to the process of composite photography, in which the same frame of film is repeatedly exposed to different scenes.[30] So long as they're close enough in form, mental images in succession generate a fuzzy general image. Each time you see an oak tree, for instance, it calls to mind all the other times you've seen one, and this new image is superimposed on the rest, producing a composite picture you think of as "oak tree."

Neither Russell nor Semon saw the contradiction between mnemic homophony and the engram. As material structures, engrams cannot simply blend into each other to form vague composites. While mental images may exhibit vagueness or fuzziness, matter always conforms to the principle of identity: x = x. An object is exactly itself, no more and no less. A vague object would lose this exact relation, being only somewhat itself and somewhat not. Of course, composite photographs always look a little fuzzy, but the picture itself, as a material object, cannot help but be precisely itself, its "fuzziness" solely in our interpretation of the picture.

The coexistence of successive perceptions in a single generalized perception cannot give the brain the ability to construct generalized bits of gray matter. Mnemic homophony was Russell's worst fear realized, for it revealed where mind fails to fit into matter.

Semon and Russell's resistance to the irretrievably immaterial nature of mnemic homophony placed them at odds with modern physics. Why base a theory of reality on matter when matter turns out to be some kind of space-stuff called *fields*? Descartes's reduction of causation to contact mechanics went out with Newton, a fact confirmed in the nineteenth century with the onset of electromagnetic field theory.

Mnemic homophony gives us memory without the need for neural engravings. Semon always thinks of Capri when he smells a particular cooking oil because he once happened to catch a whiff of it from a nearby restaurant as he gazed at Capri across the Bay of Naples.[31] No information storage is required, only the principle that any given mental state is influenced by similar previous states. Rather than encoding information about past events, the memory trace only matches whatever pattern of synaptic transmission took place during the original event, serving as a marker or sign that facilitates recall, re-establishing in consciousness a sense of the prior event much as radio antennae monitor the long-range influence of electromagnetism.

Recent neural research confirms that memory involves similarity between past and present patterns of brain activity. During the act of learning, neurons establish connections with each other. When we remember the moment the learning took place, something like the original pattern of connections is reinstated. However, as University of California researcher Jeff Johnson reports in *Neuron*, reinstatement of prior neural patterns takes place even when recall is limited to the learned information itself, without any details about the moment it was acquired.[32]

Like Russell, Johnson wants to know how the brain accounts for our sense of pastness. Since reinstatement applies not only to memory but knowledge, which is devoid of any sense of the past, neural similarity alone can't provide an answer. If recall is more than just synaptic rearrangement but the actual revitalization of past experience, the problem dissolves. We sense a depth to time precisely in the act of plumbing it.

Recall is often a struggle. Instead of arriving all at once, the memory creeps in. First we get the general sense of it, and gradually the details emerge like the tissues of an embryonic organ. Though not at all what we'd expect from a data storage system, this is exactly what we'd expect from a tuning system. The signal is first captured and then strengthened.

Whereas episodic memory involves conscious recall, habit-memory is the unconscious cumulative effect of past behaviors on current behavior. Semon illustrates the role of mnemic homophony in habit with a game of fetch. Each time his owner cocks his arm, the dog understands he's going to throw the stick. Even if he doesn't actually toss it but only pretends, the dog chases the chimera because his owner's gesture has awakened its memory of when he actually did toss the stick. Of course, this works only so many times. Before long the dog refuses to run until it has *perfect* homophony between the new stimulus and the old stimulus, i.e. when it actually sees the stick emerge from the hand. Habitual behaviors are activated by mnemic homophony, whether rough or perfect, between current and past circumstances.[33]

When musician Kristin Hersh and her band recorded for the first time in a "fancy" studio, she found herself unable to reproduce her usual vocal intensity because, as she explained to the engineer, when she performed live or in her usual practice space, she was relaxed enough to let go and allow the song to sing itself. In contrast to the song's voice, her own voice was self-conscious and forced. The recording engineer's first adjustment was to remove her voice from her headphones so she wouldn't be screaming in her ears, but the intensity of the song's voice remained elusive. What finally worked was simply to let her play guitar while singing, completing the homophony of her current performance with the abstracted essence of prior performances.

Only then did *Throwing Muses* roar to life.[34]

We all know we usually have to repeat a newly learned procedure before it becomes "automatic." But if the instruction is inscribed on neural tissue, why isn't once enough? Stored information is a digital phenomenon; the data's either encoded or it's not. Semon's memory is analogue, each performance of a procedure increasing the odds of it coming to mind with the relevant stimulus.

Like Darwin before him, Semon found the idea of evolution implausible without the ability of organisms to inherit and build upon the behavioral and bodily modifications implemented by forerunners.[35] Otherwise, ongoing adaptations to changing conditions play no role in evolution. This is why, in *The Mneme*, he reports on salamanders coaxed into either holding their young in utero longer than usual or releasing them early, in both cases their progeny carrying on the newly-altered behavior.[36] He also reports on trees transplanted from temperate to tropical regions and vice versa, either way their new adaptations cropping up in offspring.[37] Echoing Darwin's observations on farm animals, he observes that praying mantis populations grow more tame with each generation in captivity despite the absence of selection for this trait.[38]

Austrian theorist August Weismann tried to refute a plethora of such claims by cutting off the tails of hundreds of mice and noting the continued growth of tails in their offspring.[39] Yet experiments demonstrating inheritance of acquired traits succeeded precisely because researchers induced organisms to make the changes themselves, just as the environment, rather than mechanically imposing new behaviors, prods creatures into actively adapting.

This debate has long since been superseded by the sheer weight of evidence. We now know that when fertilizers tinker with the growth cycle of a crop, the new pattern of growth continues appearing for generations.[40] Defensive spines built up by *Daphnia* water flea in the vicinity of predators continue emerging in offspring never exposed to this threat.[41] A Dutch study has found reduced lifespan among people whose grandparents, in their youth, gorged themselves during rare seasons of overabundance.[42]

The question is no longer *whether* adaptations are inherited but *how*. Since none of these examples involve genetic changes, biologists refer to the phenomenon as "epigenetic inheritance," whereby newly acquired traits are passed on via modifications of chromosomes or even cytoplasm. Semon's belief that migrating engrams transmit traits by altering germ cells may not be so far-fetched after all. But mnemic homophony gives us another option. If past and present can be connected on the basis of similarity within an individual lifespan, why not across generations as well?

You would never suspect, reading Russell, that Semon insists on the inheritance of adaptations or that he denies the reduction of memory to a machine-like process. With his "law of ecphory," Semon contends that in contrast to machinery, which requires a complete input to produce a complete output, a memory can be fully realized even when the trigger, such as the smell of cooking oil, contains only a hint of the original event.[43] He notes that embryos, again in stark contrast to machines, can weather "large and arbitrary subtractions" of their tissues and resume normal development as if nothing happened.[44]

Russell was too committed to establishing Semon's materialist credentials to notice where he and Bergson overlapped. Many years later the task of synthesizing Semon and Bergson fell to a young biologist in training at Cambridge University, a theoretical nonconformist who took a year off from his laboratory work to study philosophy at Harvard. Unlike Russell, whose reading of Bergson was colored by professional rivalry, Rupert Sheldrake was captivated by Bergson's radical take on time and its implication for memory. By coupling Bergson's enduring present with

Semon's mnemic homophony, Sheldrake obtained the basis for a scientific theory of mentality, the very prize Russell sought in his *Analysis of Mind*.

Designed to explain organic development from egg to maturity, Sheldrake's theory of morphic resonance is based on his Bergsonian reading of Semon. Where "mnemic" emphasizes the emergence of organic form as a memory-based process, Sheldrake's use of "morphic" turns it the other way around, highlighting the proposition that nature's inherent memory operates on the basis of form. The more similar a current organic form to a previous form, the more it resonates with that form.

Sheldrake extended the mnemic principle beyond the brain to the whole organism, including all levels of structure comprising it, such that every organ, every tissue, every cell and organelle reproduces the actions it undertook in previous similar situations. The body-memory that maintains the adult on the basis of its personal past is no different, fundamentally, from the species-memory that guides embryogenesis.

In applying morphic resonance to the embryo, Sheldrake reconfigured memory into a property of species as much as individuals. Thus human embryos develop along the same lines as previous human embryos whereas chimpanzee eggs divide and grow along the lines of previous chimpanzee eggs. Like reciting text from memory, at each passage the embryo simply replicates the actions of its ancestors when they reached that stage. Just as a recollection is associated with a neural memory trace, development from the egg is routed correctly via genetic markers. In neither case, whether neural or genetic, does the marker contain the memory itself.

Morphic resonance is revealed wherever successive generations of a given species improve at a given task without guidance from their parents. The best-documented spontaneous case of this phenomenon concerns birds in Western Europe that learned to open milk bottles. The technique was first observed in 1921 in

Southampton, England among blue tits and spread primarily through simple imitation. However, since blue tits rarely travel more than a few miles, it's unlikely imitation could account for the appearance of this habit in Sweden, Denmark and Holland. "The Dutch records are particularly interesting," writes Sheldrake. "Milk bottles practically disappeared during the war, and became reasonably common again only in 1947 or 1948. Few if any tits that had learned the habit before the war could have survived to this date, but nevertheless attacks on bottles began again rapidly."[45]

Of course, postwar birds may have learned the process again from scratch. For this reason William McDougall's experiment on learning in rats provides a more compelling example. One of many scientists around the turn of the twentieth century to have demonstrated the inheritance of acquired traits, McDougall placed rats in a water maze and found that each generation solved the maze more quickly than its predecessor. Like Semon, he assumed the animals' genes were somehow incorporating and transmitting the acquired ability. But when the experiment was replicated, first in England and then Australia with rats unrelated to McDougall's, the tendency for improvement continued as before, an outcome inexplicable except in light of species-memory via morphic resonance.[46]

Long-range memory has also been revealed in tests on human subjects, for instance non-Japanese speakers who were better able to memorize authentic Japanese nursery rhymes than rearranged nonsensical versions.[47] According to Sheldrake this result follows from the fact that untold millions of people have already learned the rhymes, and anyone trying to memorize the correct versions is influenced by their cumulative experience. When subjects of another experiment were shown Persian words for ten seconds, some real and some only Persian-like fakes, and then asked to recall the words, they fared significantly better at reproducing the real words.[48]

Flabbergasted by Sheldrake's audacious proposal, neuroscientist Steven Rose designed an experiment that would surely dispose of it once and for all. The experiment involved day-old chicks divided into two groups, test chicks allowed to peck at yellow diodes and control chicks that pecked at chrome beads. After pecking the diodes, the test chicks were injected with lithium chloride, a toxic substance that made them mildly nauseous, while control chicks were injected with a harmless saline solution. The same procedure was followed for 37 days with a new batch of chicks each day. The data indicated that successive batches of test chicks became gradually more hesitant to peck relative to control chicks.

While this finding indicated the influence of previous experience, the most compelling result concerned control chicks given the choice of pecking at either the yellow diodes or the chrome beads. Over the course of the experiment, successive batches of these chicks became increasingly reluctant to peck at the diodes, suggesting that they were influenced by the cumulative experience of chicks that had pecked at the diodes and then been injected with lithium chloride. After stalling for months, Rose reneged on his agreement to write up the results with Sheldrake for publication.[49]

Needless to say, a handful of anecdotes and unrepeated experiments falls short of proof. While interesting, Sheldrake's theory remains largely untested. But at least it has the potential to explain development from the egg. The same cannot be said of the quaint notion that DNA is a blueprint or recipe for building an organism.

Around the time he was mutilating mice in a misguided effort to refute the inheritance of adaptations, August Weismann proposed that organisms develop from the egg on the basis of information transmitted from parents via "determinants" (now known as genes).[50] Though subsequent research seemed to confirm this idea, the gains in molecular biology that fleshed out

Weismann's theory would ultimately abolish it.

A theory is scientific insofar as it reduces a complex phenomenon, such as the organization of a living body, to something simple like information stored in DNA. At the core of Weismann's proposal was the assumption that genes are relatively simple static structures that generate the developmental machinery which, in turn, produces the immensely complicated systems that comprise the organism.[51] Different species are differently formed because each kind has a unique set of genes and therefore a unique developmental pathway.

Neither Weismann nor any of his intellectual descendants anticipated that developmental or "homeobox" genes would turn out to be virtually identical in species ranging from insects to people. As we learn from the field of "evo-devo," what changes in the course of evolution is not so much the genes themselves but the regulatory DNA that switches them on and off to ensure that development is species-appropriate.

Usually adjacent to the homeobox genes they regulate, epigenetic tags or "switches" operate at blinding speed. According to molecular biologist Sean Carroll, typical developmental processes involve "tens of thousands of switches being thrown in sequence and in parallel."[52] The operation of switches is so complex that they can be analyzed only with combinatorial logic. "Because the combination of inputs determines the output of a switch, and the potential combinations of inputs increase exponentially with each additional input, the potential outputs of switches are virtually endless."[53] Every switch position and associated pattern of protein production is but a snapshot, a single frame in "one hell of a movie with nonstop action."[54]

Imagine a forest overflowing with lightning bugs except that this forest is actually *produced* by the incomprehensibly complex and ever changing patterns of lightning bug flashes. Altering this pattern alters the shape of the forest. This, according to molecular biology, is how our bodies develop.

Whether we're looking at cycling networks of proteins in a cell or webs of feedback loops governing everything from immune response to patterns of neurotransmission, the number of possible outcomes stemming from any given input is virtually infinite, blocking the way to successful physical analysis. Genes were supposed to be the exception, something we could bring within our orbit of comprehension. Now we find that the computation of genetic activity escalates infinitely, leaving us with the absurdity of reducing one complexity to another.

But let's assume, for the sake of argument, that a given set of complex genetic operations does lead, in a purely mechanical fashion, to a given bodily form. The problem here is that we've only pushed the question back a step: what gives rise to the complex pattern of gene activity in the first place? We're back to morphic resonance, except that now, instead of newly developing organs resonating with previous organs under similar conditions, current genetic expression resonates with prior genetic expression. Whether or not the whole reduces to the gene, the organism is still explained by resonance and not genetics. It's a reduction alright but to the past rather than the small.

In light of mnemic reduction, there's no longer a compelling reason to characterize the organism in terms of its genes. Instead, both gene expression and organ development are informed by similar past activities. Rather than construct higher-level structure, the genetic level does just what it appears to do, pumping out the proteins required by cells to carry out their tasks. That certain proteins are necessary for the emergence of certain phenotypic traits in no way implies gene-protein sufficiency in the shaping of the organism.

Despite having captivated generations of biologists, Weismann's proposal has no potential as an explanatory theory. Sheldrake, on the other hand, reduces the body's stupendous complexity to an elementary property of nature, a kind of inertia of organic form. With the demise of the DNA-based theory, morphic resonance is the only game in town.

Since we don't feel like machines, it's odd that Russell had such faith in the reduction of organisms to mechanized assemblages of atoms. The most compelling data in opposition to this belief are generated daily by that ongoing half-baked experiment we call life. Unlike materialism, the mnemic theory makes room for the mind as a thing in itself, the seat of self-existence. We appear to be thinking, feeling, freely acting people — and not genetically programmed organic robots — because we are in fact people leading meaningful lives.

Russell clung to materialism like a child to his mother. By contrast Bergson and Sheldrake realized it's precisely *against* matter that memory is defined. With memory freed from the smothering embrace of matter, mind is at last made sensible.

So long as it's restricted to the brain, the mind can be dismissed as mere shadow play. Only when extended throughout the body does it find its home. Mentality is associated with every organ, guiding development and maintaining form via resonance with similar previous forms. The brain differs from other organs only insofar as it's attached to sense organs and therefore involves awareness. Where brain-mind is at least partly conscious, gut-mind operates entirely in the dark.

"Mind and body" is more phrase than reality. We have two words for the same thing because we see body-mind from two perspectives, one in terms of space and the other in terms of time. As the body is the spatialized surface of the mind, the mind is the temporal depths of the body. Accordingly, death is where the body loses its mind, where matter and memory cease to be united.

What the ancients called soul or spirit has been translated in modern consciousness as the *immaterial* element of life. But we don't have to define organic memory in the negative, any more than body-mind must be defined as the *unconscious*. The immaterial element is simply the influence of the remote past on the present. Previous actions undertaken in situations most resembling the current situation are the ones most likely to materialize.

The abstract image of "oak tree" in human thought is only a faded reflection of the deeper biological process whereby former explications of growing oaks overlap into a developmental map accessible to every sprouting acorn. Whereas the individual mind is the seat of imagination, species-mind is the seat of living formation.

Wedded to the dual reduction of the world to tangible matter and timeless law, Russell missed the message of the mind, which is neither one nor the other. In the end he got it wrong because he just had to be right.

1. Russell, Bertrand, *The Analysis of Mind*, London: George Allen & Unwin, 1921, p 92
2. *Ibid*, p 279
3. *Ibid*, p 300
4. *Ibid*, p 110
5. *Ibid*, p 89
6. *Ibid*, p 63
7. *Ibid*, p 64
8. *Ibid*, p 255
9. *Ibid*, p 285
10. *Ibid*, p 105
11. *Ibid*, p 130
12. *Ibid*, pp 283-284
13. Bergson, Henri, *Matter and Memory*, London: Swan Sonnenschein, 1911, pp 318-319
14. Russell, p 159
15. Bergson, p 20
16. *Ibid*, p 11
17. *Ibid*, p 193

18. Bergson, Henri, *The Creative Mind*, New York: Philosophical Library, 1946, pp 180-181
19. Russell, p 86
20. *Ibid*, p 77
21. *Ibid*, p 91
22. *Ibid*, pp 79-83
23. Semon, Richard, *The Mneme*, London: George Allen & Unwin, 1921, p 275
24. Russell, p 88
25. *Ibid*, p 90
26. Semon, p 240
27. *Ibid*, p 14
28. McCrone, John, 'Not-so total recall,' *New Scientist*, May 3, 2003, p 27
29. Russell, pp 218-219
30. Semon, p 164
31. *Ibid*, p 92
32. Johnson, Jeffrey D, et al, 'Recollection, Familiarity, and Cortical Reinstatement: A Multivoxel Pattern Analysis,' *Neuron*, 63, 2009, pp 697-708
33. Semon, p 156
34. Hersh, Kristin, *Rat Girl*, New York: Penguin, 2010, pp 288-291, 308-310
35. Semon, p 290
36. *Ibid*, pp 58-60
37. *Ibid*, p 64
38. *Ibid*, p 133
39. Gould, Stephen Jay, *The Structure of Evolutionary Theory*, Cambridge: Belknap Press, 2002, p 201
40. Durrant, Alan , 'The association of induced changes in flax,' *Heredity*, 32, 1974, pp 133-143
41. Young, Emma, 'Rewriting Darwin: the new non-genetic inheritance,' *New Scientist*, July 9, 2008, pp 28-33
42. Cloud, John, "Why DNA Isn't Your Destiny," *Time*, January

18, 2010, p 50
43. Semon, p 124
44. *Ibid*, p 177
45. Sheldrake, Rupert, *The Presence of the Past*, New York: Times Books, 1988, p 178
46. *Ibid*, p 175
47. *Ibid*, p 189-190
48. *Ibid*, p 192
49. Sheldrake, Rupert, 'An Experimental Test of the Hypothesis of Formative Causation,' Rivista di Biologia - Biology Forum, 86, 1992, pp 431-44. Available from: http://www.sheldrake.org/Articles&Papers/papers/morphic/formative.html (Accessed Jan 2 2014)
50. Gould, p 207
51. Bertalanffy, Ludwig, *Modern Theories of Development: An Introduction to Theoretical Biology*, London: Oxford, 1933, pp 32-33
52. Carroll, Sean B, *Endless Forms Most Beautiful*, New York: W.W. Norton and Company, 2005, p 114
53. *Ibid*, p 124
54. *Ibid*, p 128

Chapter 5

What Would Darwin Say

On a winter night in Montana's Bitterroot Valley, 200 people gathered at the junior high gym for a school board meeting. The star attraction, minister Curtis Brickley, wowed the audience with his multimedia extravaganza purporting to demonstrate that life on earth could only have been crafted by an "intelligent designer." According to Jenny Johnson, writing for the *Ravalli Republic,* Brickley "questioned the process of natural selection at the molecular level" as well as "amino acids organizing into proteins with significant biological consequences, without the presence of a designer."[1]

Creationism is a protest against the deification of DNA, both as engine of species-formation and living template in whose image we are cast. What's disputed isn't so much particular findings of science as the idea that "Darwinism" is all there is and all we are. The good reverend Brickley doesn't deny that from a natural standpoint the body's exquisitely integrated self-regulating patterned dynamics rely on the proper assembly of proteins on the basis of linear sequences of nucleic acids. He just thinks such a skimpy setup might need a *helping hand* to flesh it out.

Though the evolution debate is generally framed in stark terms — God versus Darwin, religion versus science, superstition versus reason — in reality the enemy camps represent different wings of the same school of thought. With its mechanistic ideology, neo-Darwinism weakens the case for evolution and leaves the field clear for the return of God, Master Mechanic, to the science classroom.

In the age of Newton, when the cosmic machine whirred and hummed with God-given efficiency, the apex of mechanical achievement was the miniaturized timepiece, a tribute to the power of human intelligence to impose order onto dumb matter, much as divine intelligence generated living creatures from water and sand. When theologian William Paley likened the organism to a watch, his intention wasn't to diminish the glory of living things so much as augmenting that of the Almighty, now portrayed not only as a loving father but a skilled mechanic laboring to realize a pre-envisioned plan.

But if the machines we craft here on earth can run on auto-pilot, why not the entire universe? Even if stars and planets are the handiwork of a deity, once we've cracked the code, we don't need the programmer. In the celestial logic circuits, divinity doesn't compute. God turns out to be a ghost in his own machine.

So long as life is conditioned by external universal forces, it doesn't matter whether the design process is purposive and intelligent or blind and random. Either way the organism plays no role in how it lives or the evolution of its kind.

Where scientists ask how supernatural intelligence can be put to the test, creationists ask how simple chemistry, even in conjunction with environmental selection, can yield the most complex well-formed purpose-driven systems in the known universe. Are we seriously to believe that the defining characteristics of all the species on earth, in all their splendor, precision and complexity, ultimately derive from random errors cropping up in the replication of DNA?

As Nobel-winning biologist Christiane Nűsslein-Volhard says, "the variations Charles Darwin recognized as the basis for evolution are largely caused by random mistakes during the process of DNA replication."[2] It's significant that an evo-devo pioneer like Nűsslein-Volhard would make such a claim. So ingrained is the reductionist outlook that biologists assume the source of genetic regulation is the genes themselves being regulated.

If Noam Chomsky is right that the knack for grammatical language is instinctive rather than taught, the implication for neo-Darwinism is that all the physiological and neurological changes required for the emergence of complex language were triggered by random mutations. Purely by accident the mutation enabling us to comprehend speech appeared at the very moment another mutation gave us the ability to articulate it. Isn't it more likely the motive force for human language was the desire of our ancestors to better understand each other?

Though unaware of it, when we encounter someone in person our gaze favors the right side of the face, which reveals emotion more clearly than the left side. Amazingly, dogs do this too when looking at people. Since the shift of the eye is too slight to be noticed by other dogs, the habit cannot be learned and is therefore instinctive.[3] Yet dogs evolved from wolves only in the last 30,000 years, not a very long time for an exceedingly improbable genetic accident to bequeath such a precise and useful instinct. Far more plausible is that dogs learned this trick, like many others, because they wanted to be our pals, and knowing how we feel and what we want is essential in pleasing us.

Just as the formation of the first hydrogen atom increased the odds of others forming, each time a dog learned how to read a human face, the likelihood increased that more dogs would pick up the trait, and so on, until it was embedded in canine memory, explicated by every pup.

Given the fact that camels commonly have to kneel in the desert, it's no surprise the skin over their knees is callused. What is surprising, at least for neo-Darwinism, is that calluses begin forming in the embryo and therefore arise biologically, not just through day to day pressure on the skin.[4] Supposedly the camel genome underwent a random mutation that not only conferred callused skin but specified where the rougher skin would be located — *and nowhere else.* So stupendously unlikely is such a perfectly tailored mutation that calculating the odds against it exceeds our imaginative capacity. It is, in effect, a miracle.

Preoccupied with the creeping threat of divine intervention in the sciences, the pundits of protein can't see the dependence on the miraculous in their own favored theory. Without actually explaining anything, the neo-Darwinist simply replaces the heavenly weaver stitching legs onto fish with a genetic hiccup. Like a toy in a cereal box, every defining trait of every species comes with a special mutation hidden inside. Genes, you might say, work in mysterious ways. We don't know why the right mutation comes along at the right moment — it just does!

Faced with the choice between intelligent design and its demented cousin, is it any wonder so many people prefer the former? As long as living creatures are seen as devices constructed according to blueprints — a soul deposited by God, a code written by chance — we're always tempted to revert to the cosmic mechanic, who alone seems able to account for not only the astounding intricacy of organisms but our sense of life that it *lives,* that it strives to bloom and then bask in its brilliance. The neo-Darwinian mistake is to latch onto a worldview rooted in Paley's theology and try to hammer it into an evolutionary framework. Far better to grant evolution the status of first principle, in effect deriving physics from biology. As philosopher Alfred North Whitehead put it, "biology is the study of the larger organisms; whereas physics is the study of the smaller organisms."[5]

As it happens, the whole debate was superseded long ago by an amateur naturalist who devised a theory of speciation not only rigorously scientific but compelling and believable to boot. Known as "descent with modification by natural selection," it was the product of decades of research and reflection.

"Throw up a handful of feathers," says Charles Darwin, author of *The Origin of Species,* "and all fall to the ground according to definite laws; but how simple is the problem where each shall fall compared to that of the action and reaction of the innumerable plants and animals which have determined, in the course of centuries, the proportional numbers and kinds of trees now growing on [Native American] ruins!"[6] When it comes to complex issues such as ecology, evolution and animal behavior, the mathematical abstractions of physics have little to offer. Dissenting from T.H. Huxley's concept of animal automatism, Darwin stresses the importance of individual will in shaping our actions.[7] He further concludes that delicately balanced arrays of cells, tissues and organs cannot properly function without a "coordinating power" that brings "the parts into harmony with each other."[8]

Yet today students are taught that evolution is a purely automated interplay of natural selection and random mutation. Richard Dawkins goes so far as to recast evolution as a competition between genes rather than the organisms that carry them.

In an article for *Natural History*, "The Illusion of Design," Dawkins characterizes "Darwin's idea" as "nonrandom survival of randomly varying hereditary instructions for building embryos."[9] Yet it was Weismann who reduced evolution to natural selection of random genetic variation. Not only was this not Darwin's idea but he explicitly rejects it in *The Origin of Species*. Evolution, he says, occurs "through the natural selection of numerous, slight, favorable variations; aided in an important manner by the inherited effects of the use and disuse of parts; and in an unimportant manner ... by variations which seem to us in our ignorance to arise spontaneously."[10]

According to Dawkins, "the skepticism that often greets Darwin's idea is a measure of its greatness."[11] But a theory of evolution that provides no causal role for the adaptive organism is inherently implausible. Correcting Dawkins, the skepticism that often greets *Weismann's* idea is a measure of its *weakness*.

So confused is the issue that the National Center for Science Education opposed a requirement by the Texas State Board of

Education for students to consider the "strengths and weaknesses of Darwin's theories." What have we come to when a science foundation insists that students *avoid* considering the merits of a theory? *Scientific American* gives NCSE director Eugenie Scott a pat on the back for being a "loyal defender" of Darwin, as if science is about loyalty to a given theorist rather than scrutiny of every scientist's work.[12] Add to this the irony that the evolutionary theory taught in our schools isn't even Darwin's in the first place.

What Darwin actually proposed is that the chief source of variations subject to natural selection is the day-to-day struggle to survive in the face of competition and limited resources. Darwinian evolution is a model of clarity and common sense: organisms adapt to changing environmental conditions and pass on their newly acquired behaviors to their progeny. Successful adaptations become ingrained and lead to bodily changes that also get passed on to each generation until a new species emerges from the old.

Though nominally Lamarckian, Darwin disagreed with his forerunner in one crucial respect. Whereas Lamarck believed God guided the development of life toward the emergence of Homo sapiens, Darwin sought to explain evolution naturalistically, appealing neither to the intelligence of a deity nor the determinism of physics but the good sense of creatures great and small.

"I think there can be no doubt," he writes in *Origin,* "that use in our domestic animals has strengthened and enlarged certain parts, and disuse diminished them; and that such modifications are inherited."[13] He cites many examples, such as young shepherd dogs that know, without training, to avoid running at sheep,[14] domesticated chickens without fear of cats or dogs as a result of their ancestors having grown accustomed to house pets,[15] and ostriches that can't fly because their predecessors learned to stand their ground, kicking predators instead of taking flight.[16]

The meaning of evolution is that species are self-created through the act of living and adapting. Physical forces merely provide the anatomical and environmental constraints against which organisms define themselves. Regarding the origin of sea mammals, Darwin writes, "A strictly terrestrial animal, by occasionally hunting for food in shallow water, then in streams or lakes, might at last be converted into an animal so thoroughly aquatic as to brave the open ocean."[17] Due to the variability of their bone structure, the very young can alter their anatomy through newly-acquired behaviors, enabling flat-fish, for example, to push their eye sockets a little higher with each passing generation. "The tendency to distortion would no doubt be increased through the principle of inheritance."[18]

Darwin maintained that without the ability of organisms to inherit traits acquired by their forerunners, the theory of evolutionary is dead in the water. "For if each part [of the body] is liable to individual variations at all ages, and the variations tend to be inherited at a corresponding or earlier age — propositions which cannot be disputed — then the instincts and structure of the young could be slowly modified as surely as those of the adult; and both cases must stand or fall together with the whole theory of natural selection."[19]

But what if he got it wrong? He certainly seems to have slipped up with his notion of pangenesis, a mechanism that conveys adaptations from large-scale bodily structures to reproductive cells. In the 1890s Weismann argued that animals, no matter how much they alter their behaviors, cannot affect the "determinants" in their reproductive cells. If genes are the sole vehicle of hereditary information, as Weismann assumed, then living adaptations cannot be inherited, and Darwinian evolution must yield to a mechanistic alternative. The new theory, neo-Darwinism, took final form in the 1930s with the "modern synthesis" of natural selection and genetics.

Though he presented his argument as ironclad, a done deal, Weismann offered no evidence to support it. According to paleontologist Stephen Jay Gould, the sole basis of Weismann's conviction was that he couldn't imagine how hereditary information could be transferred to offspring except by way of genes. "We accept it, not because we are able to demonstrate the process in detail ... but simply because we must, because it is the only possible explanation that we can conceive."[20]

Dawkins notes correctly that the inability of creationists to imagine how the species of life could have emerged without divine assistance doesn't mean that's how it happened.[21] We might add that the inability of neo-Darwinists, including Dawkins himself, to imagine how an organism could inherit traits acquired through life-struggle in no way negates this factor in evolution.

As cell biologist Stephen Rothman points out, adaptive qualities belong to organisms not genes.[22] It's the animal that must cope with jungle, desert or ocean, not the genome tucked away in its cozy nuclear compartment. Darwinian evolution makes sense because it places the organism, not DNA, at the center of the action. Neo-Darwinism, on the other hand, is reductionist, not in the scientific sense of reducing what is mysterious to what is knowable, but in the materialist sense of reducing the organism to its components.

Based on 40 years of experience in the field, Rothman says ideological reductionism has impaired the ability of researchers to accurately assess their own data. As an example, he offers the vesicle concept of protein transport, a stupendously unwieldy model requiring 15 to 30 mechanisms to move proteins a few microns. None of the experiments cited in the literature demonstrate these mechanisms but only describe what they would look like if they do exist. Like the gene-centric models of evolution and embryogenesis, proponents of the vesicle model have never tested it, never saying, "If it's true, then such and such should happen." No matter how much contradictory evidence accrues, confidence in the so-called theory never wavers. Why? Because it's the only model that seeks to account for the movement of protein according to blind machine-like operations.[23]

Darwin was skeptical of the mechanistic idea that every cause must be both temporally and spatially contiguous to its effect. Toward the end of *The Origin of Species,* he takes Leibniz to task for alleging that Newton introduced "occult qualities and miracles into philosophy" with his theory of gravity, which endows matter with the property of action at a distance.[24] As we've known since Einstein's day, electromagnetism and gravity both act at a distance without the need for an "aether" to act as material mediator.

According to Walter Elsasser, pioneer in both quantum mechanics and geophysics, the necessity of a material medium for the transmission of organic characteristics in no way follows from principles of physics.[25] Rather than foreclose on all other theoretical options in favor of an abstract system that casts genes as a kind of bio-aether, Elsasser suggested a new theory based on our experience of being alive. We already have, he remarked, a perfectly plausible model for retrieval of the past. It's called *memory.*

Mechanistic biology overlooks the possibility that memory is exactly what it seems to be — the continued influence of the past — and that embryos call up the organic forms appropriate to their kind in exactly the same way grownups recall where they left their keys.

Summing up the conventional view, science writer Matt Ridley asserts that "Something, somewhere must be imposing a pattern of increasing detail upon the egg as it grows and develops."[26] Elsasser might have rejoined Ridley with a riddle: *where,* precisely, is the past?

Elsasser's approach, which he presented in a 1981 paper for the *Journal of Theoretical Biology,* solves several problems at once.

No need for a blueprint from which the body is mechanically constructed if the processes within and among cells stay on course by repeating what came before. Instead of following preplanned design, whether theological or genetic, the embryo mimics the developmental steps of its ancestors. And in contrast to the determinism of both creationist and neo-Darwinist ideology, natural memory lends the organism a role in the subsequent evolution of its species. Between the randomness of molecular motion and the necessity of physical law lies a probabilistic gray area in which a creature may choose to follow its species-wide memory or — if environmental conditions have changed sufficiently — to select a new course of action.[27] Elsasser's "organismic selection" is the natural counterpart of Darwin's natural selection.

Meanwhile Darwin's "coordinating power" manifested in a series of experiments conducted in the 1980s by Miroslav Hill, then director of research at the Institut de Cancérologie et d'Immunogénétique in France. Hill and his team established a line of Chinese hamster cells, called CH1, in order to isolate mutants resistant to a toxic drug called thioguanine. In keeping with the neo-Darwinian model, Hill assumed this new line of cells would harbor a small number of randomly mutated cells resistant to the toxin.[28]

Yet Hill found no mutants among CH1 cells and therefore no survivors after exposure. Seeing no reason why CH1 cells couldn't mutate under exposure to thioguanine, he divided cells into separate flasks, only one of which contained the toxic selection agent. When the cells in the selection medium were depleted, he replenished the culture from growing colonies in the parallel culture.

After three weeks of periodically seeding CH1 cells into a flask containing thioguanine, Hill was greeted with colonies of resistant cells. Not only that but with each subsequent assay, a greater proportion of newly exposed cells survived attack.

Intrigued, Hill's team embarked on further testing. To eliminate the possibility that newly exposed cells somehow benefited from direct contact with previously exposed cells, he modified the procedure and dubbed it a "serial assay." At each passage in the growth cycle, cell samples removed from the neutral medium were placed in a fresh container of selection medium instead of being mixed with surviving cells in an otherwise depleted culture. He also introduced a different selection agent, ethionine, a toxic substance never before tested on mammalian cells.

The first 24 passages yielded no ethionine-resistant survivors. Survivors began appearing with the next passage, and with each subsequent passage a higher proportion of cells survived. The more cells that had already been exposed to the toxin, the more readily newly exposed siblings adapted to the same conditions. Hill inferred that the adaptive mutation was occurring not only in cells subjected to the toxin but in cells growing in parallel cultures with no toxin.

Numerous repetitions of the serial assay, including exposure to high temperature, produced the same results. Concluding that cells under attack were somehow sharing information with siblings in separate containers, Hill wondered if the "nonseparability" he observed was related to the quantum phenomenon of entanglement.

The evolutionary advantage of nonseparability is obvious. If your relatives at a different location are under pressure from an environmental challenge, your survival is much more likely if you begin adapting before the new threat arrives.

Entanglement is known to occur only among particles, which are always identical to other particles of the same type. The mathematical description of entanglement has no application to cells since, aside from their vastly greater scale, they're never perfectly identical. Yet the resemblance of nonseparability to entanglement is undeniable. Either way the effect involves shared origin: just as a pair of entangled low-energy photons was at one time a single high-energy photon, in each of Hill's experiments all the cells were descended from the same mother cell.

Like superposition, entanglement is destroyed in the course of decoherence, the "leakage" of quantum properties into the local environment. Nonlocal correlations are swept away precisely when particles interact with their surroundings. What physics calls decoherence biology calls *death*, whereby a living system loses its coherence as its components are absorbed into the environment.

It's almost spooky, a word Einstein famously applied to entanglement. Can it be that the onset of life triggered the evolution of entanglement? Perhaps in the context of an organism, nonlocal coherence occurs on the basis of mere similarity and at macroscopic scales. If so, superposition isn't the only property drawn from the quantum ground state into the tree of life. Just as Darwin showed how a hand and a paw aren't merely analogous but homologous, biological nonseparability may be descended from quantum entanglement. Even as they follow different developmental paths, sibling cells remain correlated due to shared origins. Not only the cells within them but organisms belonging to a given species are united in their resonance with the same ancestral forms. In memory we are entangled.

It's no surprise that much of the evidence Rupert Sheldrake invokes in support of morphic resonance also bolsters Hill's evolved form of nonseparability. The work of twentieth-century naturalist Eugene Marais, for instance, demonstrates nonsensory coordination among termites engaged in constructing an arch. Sheldrake attributes the seemingly telepathic abilities of termites to a "morphic field." However, since all the termites of a given mound are derived from a single termite queen, we can just as easily chalk it up to nonseparability.

To build arches, the central supports in their mounds, termite populations divide in two, each group building half the arch from the ground up until the two sides meet in the center. Curious as to whether these blind insects coordinate their efforts via smell, Marais inserted a steel plate between two groups of termites. When the termites were done, the two halves of the arch met perfectly on either side of the plate.[29] The mammalian heart develops the same way, in two parts that ultimately match up like pieces of a jigsaw puzzle. Perhaps heart cells, like termites, get their bearings through nonseparability.

As we know from studies in evo-devo, genes involved in embryonic development are essentially unchanged over hundreds of millions of years. Profoundly different classes of life are instead distinguished by differences in how genes are expressed. In place of a genetic blueprint implemented by worker-proteins, we have a complex epigenetic system rapidly switching genes on and off. Nonseparability is implicit in the instantaneous coordination of protein switches across developing organs.

The phenomenon of parallel evolution, one of biology's foremost anomalies, resists explanation along the lines of neo-Darwinism, according to which the striking resemblance between placental and marsupial mammals, for instance, stems from environmental similarities, as though similar foliage and climate could cause virtually identical wolves, squirrels, moles and jerboas to emerge independently on separate continents. Far more reasonable is the supposition that adaptive information, as Hill calls it, can be shared nonlocally among related but isolated species, enabling them to evolve along parallel tracks.

It was on the basis of parallel development in the womb across radically different species that Darwin identified mammals, birds and reptiles as "the modified descendents of some ancient progenitor, which was furnished in its adult state with bronchia, a swim bladder, four fin-like limbs, and a long tail, all fitted for an aquatic life."[30] The approximate recapitulation of phylogenetic descent during embryological development provided Darwin with an important clue that his grandfather Erasmus was onto something.

For Darwin recapitulation makes perfect sense. Since natural selection operates on adaptations made by creatures after they've left the womb, embryos ought to remain roughly the same even as juvenile and adult forms evolve. But this assumes adaptations are driven by life experience and the struggle for survival. Since neo-Darwinism transfers the source of adaptation to genetic mutations, recapitulation is a big problem. Genetic accidents can happen at any point in the developmental process. Mutations that streamline embryogenesis, reducing the time and energy required for gestation, ought to be environmentally selected right along with those that beneficially alter the adult body. Do we really need a swim bladder and fins in the womb? Why can't we bypass all that and go straight to human? The persistence of recapitulation *in utero* signifies that inheritance results not so much from ever-malleable genes but species-wide memory of a past that cannot be altered.

According to embryologist Paul Weiss, the gene-centric model fails the test of thermodynamics, which recognizes indeterminism at the molecular level and therefore restricts itself to statistical analysis at the large-scale level.[31] Weiss made a habit of debunking popular beliefs about life, such as our picture of cells as tiny factories built up from definite components. In reality the cell stands upon molecular disorder. "The only thing that remains predictable amidst the erratic stirring of the molecular population of the cytoplasm and its substructures is the overall pattern of dynamics."[32] Since organisms must maintain order in the large despite disorder in the small, how could DNA molecules be the prime force for order? Rather than controlling the organism, mechanistic processes such as DNA replication are tools by which the system maintains large-scale order.[33]

Determining eye color is a long way from directing the construction of eyes. Brown eye color follows from the presence of a pigment produced by an enzyme that results from the combination of several genes. If any one of these genes is absent, so is the pigment, and the eye comes out blue instead. What does this tell us about the structure of an eye or how it works? The fact that genes promote distinguishing traits such as eye color means only that they *influence* development, altering it one way or another, not that they program and control it down to the last detail. That genes account for the many ways we differ doesn't endow them with the power to provide the overall form of a developing organism.[34]

Weiss wondered why, if genes are in charge, living systems emerge first at the level of the whole body and descend, step by step, to specific tissues.[35] Isn't this the opposite of what we would expect from a process driven from within the nuclei of our cells? And if the body follows a pre-determined program, why does embryogenesis begin indeterminately, differing from case to case, as if the embryo must improvise until entering a definite developmental pathway?[36]

As biologist Harry Rubin explains, an organic function as elementary as the timing of penicillin production in the mold Aspergilla involves so many possible gene couplings that the actual sequence of combinations is "transcalculational, or beyond the power of any conceivable computer in a finite period of time."[37] Are we to believe, then, that the genome of a mold possesses a *magical power* of computation? Isn't it more plausible that its genes simply mimic the sequence of steps long taken by the mold's predecessors? Keep in mind that penicillin is an elementary organic compound. What about the formation of a trillion-celled furry creature? Apparently DNA is the one thing that really can pull a rabbit out of a hat.

Even the very first step of forging an organism, the production of proteins, is beyond the power of genes. A protein isn't just a thread of amino acids but a complex four-fold structure. We know from physics that an amino acid chain, or polypeptide, will fold into the minimal energy state, that is, the shape with the least amount of potential tightening up. Yet a given polypeptide is faced with a huge number of minimal energy states into which it could fold. How does it wind up in the correct shape? This is known as the multiple-minimum problem. Sheldrake's answer is that the polypeptide, because it resonates with previous polypeptides of the same sequence, naturally folds up the same way they did.[38]

If memory is a general feature of nature like gravity or electromagnetism, it ought to be evident in nonliving systems. We should find it in the crystallization of not only proteins but ordinary minerals.

It's long been known that once a particular type of crystal has been manufactured, it's liable to spontaneously pop up in other laboratories. Perhaps, as some researchers have joked, fragments of newly created crystals travel from lab to lab on the beards of migrant chemists. In light of morphic resonance, however, the mystery clears up. Once a chemical compound has crystallized in a particular pattern, other similar compounds are more likely to crystallize that way as well.

As man-made crystals are repeatedly manufactured, says Sheldrake, they should become more stable as they resonate with a greater number of previously formed crystals. By contrast, natural crystals have been around so long that their forms are already well established and can't be expected to become noticeably stronger.

Crystal stability is measured by to the temperature at which it melts. The more heat required to melt a crystal, the more stable it must be. Sheldrake discovered that melting points have indeed consistently risen among artificial crystals while remaining steady for natural crystals. For instance, the natural form of cocaine has the same melting point it had a century ago, while the melting point for the recently developed artificial form has risen 13 degrees centigrade. Whereas the melting point of aspirin has increased by eight degrees since being synthesized in the late nineteenth century, that of its natural counterpart, salicin, has remained constant. Sheldrake offers many more examples.[39]

The mechanistic response to this inconvenient fact is that chemists are getting better at purifying their samples before measurement. With greater purity comes more accurate readings of melting points. But this fails to account for the lack of rise in melting point among natural crystals.

Despite the evidence, the mechanistic worldview continues crystallizing in the minds of scientists. Though Sheldrake's theory requires only that memory is a property of nature exploited by clever and resourceful creatures, biologists would rather believe that the complete blueprint of a dynamic real-time body is statically inscribed in its chromosomes, much as physicists think time is statically inscribed in a fourth dimension. Is it just another of those fantastic neo-Darwinian accidents that a mechanistic outlook reigns supreme over biology at the very moment human society is spellbound by its own mechanical creations?

Darwin's contribution wasn't just to identify life-struggle as the main trigger of evolution but to shift the focus from the transcendent abstractions of theology and mathematics to the tangible and irreducible powers of destruction and creativity that play out day by day in the natural world. "There is grandeur in this view of life, with its several powers" such as growth, reproduction, inheritance, variability and the will to survive, none of which reduce to mechanical principles.[40]

Isn't it time the Darwinian revolution came full circle? No species of life is the passive product of external forces, whether intelligent or blind. To deny the first but not the second is to leave the revolution half undone. Darwin's declaration of independence establishes freedom from any and all celestial proclamations regardless of where they originate – in a book or with a bang – so long as they defy our birthright of active self-creation.

As Elsasser and Sheldrake demonstrate, with a bit of imagination we can establish a basis for the inheritance of adaptations and thereby escape the sterile clash of creationism and neo-Darwinism. Not only can we conceive of life liberated from the control of a Mechanic or a bundle of blind mechanisms but we have no choice, as the ghost of mechanism past will surely haunt us until we've expelled it in both its guises.

1. Johnson, Jenny, "Intelligent design presentation draws hundreds," *Ravalli Republic,* December 12, 2003
2. Nűsslein-Volhard, Christiane, *Coming to Life: How Genes Drive Development,* Carlsbad, CA: Kales Press, 2006, p 36
3. "Dogs Decoded," *Nova,* November 12, 2011. Available from: http://www.pbs.org/wgbh/nova/nature/dogs-decoded.html (Accessed Jan 2 2014)
4. Sheldrake, Rupert, *The Presence of the Past,* New York: Times Books, 1988, p 140
5. Whitehead, Alfred North, *Science and the Modern World,* Simon and Schuster, 1997 (1925), p 103
6. Darwin, Charles, *The Origin of Species,* New York: Modern Library, 1993 (1859) pp 102-103
7. Gillespie, Neal, *Charles Darwin and the Problem of Creation,* Chicago: University of Chicago Press, 1979, p 141
8. Darwin, Charles, *The Variation of Animals and Plants Under Domestication,* London: Murray, 1875, vol. 2, p 354
9. Dawkins, Richard, "The Illusion of Design," *Natural History,* Vol. 114, No. 9, November 2005, p 36
10. Darwin, 1993, pp 636-637
11. Dawkins, 2005, p 36
12. Wilcox, Kate, "Scientific American 10: Eugenie Scott," *Scientific American,* June 2009, p 64
13. Darwin, 1993, p 175

14. *Ibid*, p 324
15. *Ibid*, p 326
16. Darwin, Charles, *The Origin of Species*, 2nd Edition, Oxford: Oxford University Press, 1996 (1860) p 111
17. Darwin, 1993, p 282
18. *Ibid*, pp 290-292
19. *Ibid*, p 331
20. Quoted by Gould, Stephen Jay, *The Structure of Evolutionary Theory*, Cambridge, MA: Belknap, 2002, p 202
21. Dawkins, Richard, "Paranormal or Perinormal?" (lecture) Las Vegas, Nevada, Amaz!ng 3 Conference, sponsored by the James Randi Educational Foundation, January 15, 2005
22. Rothman, Stephen, *Lessons from the Living Cell: The Limits of Reductionism*, New York: McGraw Hill, 2002, p 8
23. *Ibid*, pp 138, 176, 161, 256, 260-261
24. Darwin, 1993, pp 637-638
25. Elsasser, Walter, *Reflections on a Theory of Organisms: Holism in Biology*, Baltimore: Johns Hopkins University Press, 1998 (1987) pp 7, 140
26. Ridley, Matt, *Genome*, New York: HarperCollins, 1999, p 174
27. Elsasser, 1998, pp 5, 41-44
28. Hill, M, "Adaptive State of Mammalian Cells and Its Nonseparability Suggestive of a Quantum System," *Scripta Medica* (BRNO) V 73 (4), October, 2000, pp. 211-222
29. Marais, Eugene, *The Soul of the White Ant*, New York: Penguin, 1973, pp 119-120
30. Darwin, 1993, p 599
31. Weiss, Paul, *The Science of Life*, Mount Kisco, NY: Futura, 1973, p 20
32. *Ibid*, pp 39-40
33. *Ibid*, p 63
34. *Ibid*, p 62
35. *Ibid*, p 28
36. *Ibid*, p 21

37. Rubin, Harry, introduction to *Reflections on a Theory of Organisms,* by Walter Elsasser, Baltimore: Johns Hopkins University Press, 1998, p xiv
38. Sheldrake, Rupert, *The Presence of the Past,* New York: Times Books, 1988, pp 123-126
39. Sheldrake, Rupert, *A New Science of Life,* Third Edition, London: Icon Books, 2009, pp 251-254
40. Darwin, 1993, pp 648-649

Chapter 6

The Anti-Sheldrake Phenomenon

By devising and testing a theory of natural memory, Rupert Sheldrake established himself as the world's foremost proponent of a thoroughgoing evolutionary science. Heir to the lineage of Darwin, Peirce, Bergson, Bohm and Elsasser, he bears the weight of their worldview. Attacks on his work amount to an offensive against any alternative to a universe under the control of fixed eternal laws. Though the idea of timeless principles governing passive nature is a leftover of an age when God the Father was our heavenly lawgiver, those who denounce Sheldrake fancy themselves champions of science and modernity.

In 1980 David Bohm speculated that tangible existence is the flipside of an invisible order bound by neither space nor time. Once an organic form pops up in the everyday explicate world of matter, it's abstracted into the implicate order, increasing the likelihood that it will explicate again under similar circumstances.

In the 70s Sheldrake formulated his proposal that organic events are influenced by composites of previous similar events, i.e. morphic resonance. Wondering if he'd stumbled onto a universal feature of nature, he investigated crystallization for evidence of natural memory outside the biological domain. In a kind of mirror image, Bohm applied his theoretical physics to living phenomena, observing that every organism explicates its species, which is implicit in each of its members. From radically different starting points, each theorist encountered the possibility of mind as a special application of natural memory.

Just as an object in motion tends to stay in motion, organic form tends to endure. Call it morphic inertia.

Struck by the similarity of their findings, the quantum

physicist and the plant cytologist met up to talk shop. Were they proposing different theories overlapping in certain respects, or was this a single theory expressed in radically different terms? So perplexing was the question that they left it open.[1]

Yet their books received very different receptions. While Bohm's *Wholeness and the Implicate Order* garnered the respect owing to any scientific work, Sheldrake's *A New Science of Life,* released in 1981, evoked not just hostility but hysteria, though the initial response had been restrained, even positive in places.[2] Only after John Maddox struck the match did the anti-Sheldrake phenomenon catch fire.

While editor of *Nature,* Maddox proclaimed Sheldrake's book "the best candidate for burning there has been for many years." As he elaborated for the BBC in 1994, "Sheldrake's is not a scientific theory. Sheldrake is putting forward magic instead of science, and that can be condemned, in exactly the language that the Pope used to condemn Galileo, and for the same reasons: it is heresy."[3]

Having confused science with the unfounded metaphysical belief that reality is reducible to matter, Pope John couldn't accept that Sheldrake's proposal, which defies this primitive belief, is logically coherent and tests out across a range of phenomena. Adding to his revulsion was the fact that Sheldrake was an insider, a scientist with impeccable credentials, including numerous publications in peer reviewed journals, *Nature* itself among them.[4]

Whether or not morphic resonance is real, that Sheldrake proposed a scientific theory is a simple fact. By denying this fact, Maddox sinned against science. Perhaps dimly aware of his betrayal, he projected his heresy onto Sheldrake rather than face up to his own failure.

Materialist biology may think it's got physics backing it up, but physics isn't so sure. According to Gabor Forgacs and Stuart A Newman, authors of *Biological Physics of the Developing Embryo,* the recognition that organisms cannot disobey physics "is of limited value" in explaining their behavior. While physics operates by reducing complex systems to simple components, in organisms complexity is a "fundamental and irreducible property." Because complexity must be taken "as is," physical analysis explains only so much when it comes to living systems.[5] Rather than the molecule, the central actor is the system of organization as a whole, i.e. the organism. Left unchecked, a physical process that aids development ultimately derails it. The timely initiation and termination of mechanical processes within and among cells, say Forgacs and Newman, "ensures that each driving force is constrained ... and that the whole complex of forces is subordinated to the survival and propagation of the organism."[6]

With numerous possible mechanisms to employ in the act of putting itself together, the organism is continually faced with choices, some routine, others life or death. How does it choose? Forgacs and Newman follow the standard assumption that embryonic development is guided by information encoded in its genes. The basic idea can be expressed by a simple equation: physics + genes = organism.

But how does the embryo know which genes to activate at any given point in its development? Since the answer cannot be the genes themselves, molecular biologists suggest that a cell's position in the embryo determines its developmental path. Simply by occupying a particular spot in the blastula, a given blastomere is destined to turn out as a certain kind of differentiated cell. All it takes is non-uniform distribution of chemicals known as morphogens, and voila, all the instructions are automatically in place for determining which genes to turn on at which places and times in order to construct the completed organism. Morphic resonance offers a less miraculous solution: the embryo simply does whatever its forerunners did when they reached the same developmental stage.

After identifying plants that inherited mutant DNA from their parents yet displayed the wild type trait belonging to their grandparents, researchers assumed a backup copy of the grand-parents' gene was contained in the plants' RNA. *The Scientist* went so far as to report that RNA inheritance had been established, later apologizing to readers for the error and noting that researchers had established only the existence of non-DNA inheritance.[7] Seems that even materialism exhibits morphic inertia.

How else to explain the resilience of the materialist assumption in the face of the facts? It's not as if the laws by which atoms behave are themselves built up from atoms. If timeless principles are to be allowed, why not mentality? Why not memory as a temporal analogue of action at a distance?

Despite the weakness of the materialist model, the Maddox editorial rendered Sheldrake taboo, and whoever touches taboo, as Freud observed, becomes taboo. The more professionally advanced, the more a scientist stands to lose by accepting the explanatory potential of Sheldrake's proposal. In such an atmosphere bullies operate with impunity. Demonstrating no real knowledge of Sheldrake's work and studiously ignoring the actual reasoning behind his theory, ideologues have assaulted and demonized him without having to face the consequences that ordinarily follow from such behavior.

"Science," according to renowned theorist Richard Feynman, "is the organized skepticism in the reliability of expert opinion."[8] Yet *Skeptic* magazine, edited by *Scientific American* columnist Michael Shermer, is all about fringe beliefs and rarely takes on expert opinion. Like his forerunner Martin Gardner, Shermer serves as an enforcer of establishment beliefs, not a critic.

Shermer seems to think science is a popularity contest. "The person making the extraordinary claim," he writes, "has the burden of proving to the experts and the community at large that his or her belief has more validity than the one almost everyone else accepts."[9] What Shermer overlooks is how hard it is to unlearn a deeply ingrained habit of thought reinforced at the group level both in the explicate sense of peer pressure and the implicate sense of numerous individuals unconsciously resonating with like-minded predecessors.

Though Shermer calls attention to the fallacy of *post hoc ergo propter hoc* — after an event therefore because of it — he doesn't realize this encapsulates the entire approach of materialistic biology.[10] After genes are transcribed, organs and tissues form. Ergo gene transcription causes bodily formation. It doesn't occur to orthodox biologists that both events could result from a deeper cause, namely the efforts of the organism to match the steps taken by its predecessors.

True believer in the reduction of the organism to machine-like operations, Shermer's prime directive seems to be the prevention of any skeptical inquiry into his own ingrained belief. He revealed this attitude during a 2007 interview when I asked him what he thought about the creativity of living things.

> What does this mean, internal creativity? Positing some sort of metaphysical thing or something. At some point we're gonna ask, well, what is this internal creativity? Quit using that word. Give us something that we can actually test in the lab. What are you talking about — genomes? Protein chain things? What is it you're talking about?[11]

Shermer's Orwellian imperative would stifle inquiry to the point where we can't even articulate our thoughts. As he put it, "instead of speculating about some inherent force at work, let's just don't call it anything." Apparently it's okay to speculate about forces or properties of nature when you're a physicist, but in biology everything has to be broken down to something akin to the workings of a cuckoo clock.

You'd think the idea that genes and their signature proteins conspire to build a living body from carbon-rich compounds would be subject to skeptical evaluation, but Shermer obediently follows authority on the assumption that all fundamental mystery has been eliminated, whether it's how organisms emerge from eggs or how species emerge from ancestral species.

When I suggested evolution involves the innate intelligence of organisms, Shermer went right off the rails.

> This is not a debate in science. No one has this debate. I've never heard this debate before, and I go to all these evolution conferences. Nobody debates this. This is an outside of science debate. This is a Rupert Deepak Chopra debate. It's a different kind of creationism. But it has nothing to do with science.

So the idea that organisms creatively adapt their behavior, which triggers bodily changes that are passed onto offspring, is just another kind of creationism. Darwin's theory of evolution, in other words, is a kind of creationism. But there's a certain logic to Shermer's view. After all, Darwin wasn't a professional biologist, i.e. an authority figure, just a devoted student of nature. No wonder Shermer called his proposal of bears foraging in water as an early step in the evolution of sea mammals "kind of a stupid example."

Parroting the reductionist misinterpretation of self-organized complexity, according to which complex systems merely amplify the simple rule-bound behavior of their components, Shermer tells me that even complexity is just a disguised form of simplicity. "At some point," he says, "you have to have a stepwise, bottom-up, natural, self-organized complexity out of simplicity."

Quite the contrary. According to the science of nonequilibrium thermodynamics, complex systems are fundamentally holistic and goal-directed. It's not the properties of particles that count but the overall organization.[12] A tornado pops into being in the act of eliminating the temperature gradient between warm air near the ground and cold air up high. The key is the flow of energy unleashed by gradient reduction, which powers the system until the source of energy is lost, i.e. when the gradient is gone. Rather than build up from molecular simplicity — stepwise or otherwise — self-organized complex systems emerge fully formed as if from the head of a thermodynamic Zeus. DNA doesn't build organisms any more than dust builds tornados.

Shermer's attitude is encapsulated in the title of his 1997 collection, *Why People Believe Weird Things*. "Some things," he writes, "such as water dowsing, extrasensory perception and creationism, have been tested and failed the test often enough that we can provisionally conclude that they are false."[13] He's wrong on all counts, though for different reasons.

To include creationism here makes no sense, as this belief represents a rejection of science and, unlike scientific propositions, cannot be put to the test and definitively refuted.

Water dowsing is indeed testable, though results are ambiguous. Some practitioners find underground water sources at rates no better than chance, while others succeed astonishingly well. Einstein saw dowsing as evidence that the human nervous system responds to "certain factors which are unknown to us at this time."[14] By contrast, Shermer dismisses it without so much as a glance at the evidence, thereby aligning himself with the Inquisition, which condemned "water witching" for precisely the same reason, because such an inexplicable practice introduces uncertainty and defies the orderliness of the reigning belief system, whether based around God's will or mathematical law.

Telepathy, on the other hand, has been verified countless times in carefully controlled laboratory conditions. Such experiments are a matter of public record, and their significance has been confirmed by independent statisticians. Shermer could read all about it in Dean Radin's *The Conscious Universe* or Chris Carter's

Parapsychology and the Skeptics. Instead he rejects it without investigation, as if all "psychic" entanglement is simply impossible. If physicists had taken that approach with quantum entanglement, we'd have lost an important lesson about nature.

In keeping with his blind faith in expert opinion, Shermer says scientists admit to their errors while "pseudoscientists ... ignore or rationalize failures." This tendency is known as "heads I win, tails you lose."[15] An experiment that demonstrates the sought-after results is legitimate, but if it shows negative results, it must have been poorly designed or executed.

As Stephen Rothman reveals, this mindset is alive and well in establishment science. For many years an esteemed researcher at UC San Francisco, he raises numerous examples of "heads I win, tails you lose" trickery among his colleagues, in particular regarding the vesicle model of protein transport, which posits that proteins can't spontaneously go where they're needed in a cell but must be transported by a mechanism. If protein is found where the model predicts, it verifies the model. If protein is found in the wrong location, it's written off as a "contaminant." In the use of autoradiography, scientists determine the best exposure time according to whichever time produces the results they were looking for. If samples fail to show the desired results, it's assumed the process of taking samples was somehow flawed. And so on.[16]

The basic problem, says Rothman, is the failure of most experts to doubt their own beliefs. It doesn't help that "skeptics" encourage them in their self-certainty. "The call to authority," he warns, "is sometimes dressed in the garb of scientific skepticism."[17]

Robert Todd Carroll, author of *Skeptic's Dictionary*, accuses Sheldrake of the unforgivable crime of working outside "scientific paradigms."[18] Yet Sheldrake's central proposal, that natural systems are influenced by similar previous systems, can easily be put to the test, as neuroscientist Steven Rose demonstrated when he tried (and failed) to refute it. Carroll claims that morphic resonance leaves no room for genetic influence over the forms and functions of the organism. Sheldrake, however, is absolutely clear on the need for a complementary approach, with genes differentiating individuals within a given species and morphic resonance providing the general background form common to all members of the species.[19] Compounding his confusion, Carroll asserts that Sheldrake has substituted laws of nature with morphic resonance. In fact it was philosophers CS Peirce and Alfred North Whitehead who substituted laws of nature with habits, and Sheldrake's contribution is to *explain* universal habit according to morphic resonance.

Carroll even goes so far as to associate morphic resonance with Plato's concept of a timeless realm of static Forms, ignoring the fact that Sheldrake's hypothesis is explicitly designed to allow organic forms to evolve rather than merely reflect fixed eternal types.[20]

Unable to argue his way around Sheldrake, Carroll attempts to smear him as an occultist peddling metaphysics and the paranormal. Given the well-established concept of action at a distance in physics, the occult charge is inexcusable. As to metaphysics, Sheldrake avoids it like the plague, both in the colloquial sense of supernatural explanation and the philosophical sense of a theory of reality. Carroll, meanwhile, embraces the metaphysics of mechanism. As to the paranormal, Sheldrake points out that if telepathy turns out to be real, by definition it's normal.

Summoning the old fallacy of guilt by association, Carroll attributes to morphic resonance the same scientific status as the engram, a term he claims was coined by L Ron Hubbard. In fact, the concept of the engram was introduced by Richard Semon, though it was Semon's concept of mnemic homophony that really informed Sheldrake's work.

In contrast to writers like Maddox, Shermer and Carroll,

scientists generally have enough sense to refrain from condemning the work of other scientists, particularly if they haven't taken the trouble to understand it. Alas this is not the case with biologist PZ Myers and his uninformed ad hominem attacks against Sheldrake on his blog, "Pharyngula." In his entry, "The Sheldrake Phenomenon," Myers reveals his own religious-like attitude toward nature.

Myers begins by referencing a discussion between Sheldrake and Richard Dawkins in which Dawkins, apparently unaware of quantum entanglement, asserted that telepathy would "turn the laws of physics upside down" and that demonstrating its existence would therefore require extraordinary evidence. Sheldrake responded that it would be far more extraordinary if everyone who claims to have experienced telepathy is deluded. Myers twists this statement around by claiming that Sheldrake denies the possibility that people could be deluded about their experience. On this basis, Myers pronounces his verdict. "This man is nuts."[21]

Sheldrake is nuts because he thinks a popular belief just might be true? Apparently Myers views reason as the province of an educated and trained elite. Since ordinary people can't possibly think logically on the basis of evidence, in order to know anything, they must consult a member of the scientific priesthood, like Myers for instance.

Yet Myers hasn't done his homework. His major point of dispute is that Sheldrake provides no mechanism for telepathy, and therefore his research is meaningless. Aside from the obvious fact that science routinely investigates phenomena without knowing their cause, Myers seems to have no idea that Sheldrake is first and foremost a biologist and that his interest in telepathy grew out of morphogenetic field theory, which emerged from developmental biology in the 1920s.

And it's here that Sheldrake comes in for legitimate criticism.

As biologists came to grips with the problem of how activities are coordinated within and among cells, Paul Weiss and other theorists began discussing the idea of a field effect within developing organisms. Like a magnetic field, the morphogenetic or "form-giving" field would inform cells as to their proper place in the embryo. But Weiss doesn't seem to have taken the term literally, and to this day it's generally not considered an actual physical field. Arguing that they're as real as gravitational and electromagnetic fields, Sheldrake says they coordinate activities of cells within bodies as well as insects within colonies, even enabling telepathy and the sense of being stared at.

While there's good evidence for morphic resonance, is a field concept really needed as well? True, the parts of an embryo seem to be coordinated much in the way that iron filings are brought into order by a magnetic field, but once we admit to the existence of the morphogenetic field, it becomes an intermediary between current and past organisms. Instead of resonating directly with past embryos, the current embryo is organized by a field that, in turn, resonates with past embryos. This introduces a wrinkle into an otherwise elegant theory.

Miroslav Hill's application of entanglement to biology does away with the need for a field effect. Just as photons are entangled insofar as they materialize a common form, cells of a given type may become entangled in the process of resonating with the same ancestral form. Even different kinds of cells within a body could be entangled simply by their common origin in the fertilized egg.

In his discussion of the phantom limb effect, whereby an amputee still senses a limb after its removal, Sheldrake proposes that the morphogenetic field that governed the formation of an arm remains in place even after the arm itself is lost. But if the amputee is in resonance with his own past, including the time he still had both arms, this alone would explain the phenomenon. Same goes for the mysterious but demonstrable ability of pets to accurately anticipate their owners' return from a distant location.[22] Just as entanglement is most evident among particles that recently interacted, once a pet and its owner have bonded, they too might remain nonlocally bound.

The morphogenetic field has always occupied the borderlands between reality and heuristic concept. As soon as Forgacs and Newman claim these fields are nothing more than the distributions of morphogens, they turn 180 degrees and assert that morphogens are distributed by morphogenetic fields.[23] No one seems to know how to approach the field concept, the only agreement being that the parts of embryos are coordinated in a way that has yet to be explained from a materialist standpoint.

Even if Sheldrake slipped up with his literal reading of morphogenetic fields — much as Darwin went astray with his concept of pangenesis — this in no way justifies the disgraceful treatment he's received from ideologues both inside and outside the sciences.

To really criticize Sheldrake, you've got to open your mind enough to acquire a basic understanding of his work. Like Maddox before them, Shermer, Carroll and Myers think they can dismiss one of our foremost thinkers with a few insulting claims. Incinerating other people's ideas is the quickest way to look smart. Proposing and testing a solution takes commitment and courage, all the more so if it departs radically from received wisdom.

What "scientific skeptics" reveal in their attempts to smear Sheldrake and banish morphic resonance — most recently on the pages of *Wikipedia* — is fear of discovery. They mistake their closed system of thought for science itself, as if anything at odds with their belief system is by definition anti-scientific. Like a cult, they reinforce each other's deluded sense of righteousness in the face of implacable and unreasoning enemies, all the while imagining that their efforts at maintaining collective self-satisfaction amount to some kind of noble undertaking.

Whereas true skepticism is the doubt that flushes out superstition and makes way for knowledge, cult skepticism is the absolute certainty in the falsehood of any suggestion in conflict with the ingrained prejudice of its adherents. Cult skeptics believe in a materialistic science closed off to any alternatives, much like the deterministic universe they accept without question.

Increasingly complex natural organization expresses the evolutionary pathway laid down in the implicate order. This applies to human culture as much as atoms, crystals and organisms. But there's a key difference. Whereas nature's abstractions faithfully reflect reality, the combined power of prejudice and peer pressure all too easily diverts human-devised abstraction into self-contained systems of delusion.

The world is chock-full of irrational thinking, and the methodology of science offers an antidote. For this reason the systematic confusion of science with dogmatic materialism is the most dangerous cult of all.

1. Sheldrake, Rupert, *A New Science of Life* (3rd Edition), London: Icon Books, 2009, pp 303-309
2. *Ibid*, front cover
3. Quoted from a video of the interview. Available from: http://www.youtube.com/watch?v=QcWOz1xjtsY (Accessed Jan 2, 2014)
4. Sheldrake, A.R., "The ageing, growth and death of cells," *Nature*, 250, 381-385, August 2, 1974
5. Forgacs, Gabor and Newman, Stuart A, *Biological Physics of the Developing Embryo*, Cambridge: Cambridge University Press, 2005, p 1
6. *Ibid*, p 50
7. Hrastar, Laura M., "Is RNA inheritance possible?" *The Scientist*, March 23, 2005
8. Quoted in Smolin, Lee, *The Trouble with Physics*, New York: Houghton Mifflin, 2006, p 307
9. Shermer, Michael, *Why People Believe Weird Things:*

Pseudoscience, Superstition, and other confusions of our time, New York: W.H. Freeman and Company, 1997, pp 50-51
10. *Ibid,* p 53
11. Shermer, Michael, interview with the author, "Dr. Michael Shermer on Darwin, Evolution and Creativity," *Skeptiko,* February 12, 2007. Available from: http://www.skeptiko.com/3-dr-michael-shermer-on-darwin-evolution-and-creativity/ (Accessed Jan 2, 2014)
12. Schneider, Eric and Sagan, Dorion, *Into the Cool,* Chicago: University of Chicago, 2005, p 113
13. Shermer, 1997, p 16
14. Einstein, Albert, letter to Herman Peisach, February 15, 1946. Available from: http://archive.is/LXs7v (Accessed Jan 2 2014)
15. Shermer, 1997, p 53
16. Rothman, Stephen, *Lessons from the Living Cell: The Limits of Reductionism,* New York: McGraw-Hill, 2002, pp 185, 210, 213
17. *Ibid,* p 283
18. Carroll, Robert Todd, *The Skeptic's Dictionary: A Collection of Strange Beliefs, Amusing Deceptions, and Dangerous Delusions,* Hoboken: John Wiley & Sons, 2003, pp 231-232. Available from: http://www.skepdic.com/morphicres.html (Accessed Jan 2 2014)
19. Sheldrake, Rupert, *The Presence of the Past,* New York: Times Books, 1988, p 89
20. *Ibid,* pp 106-107
21. Myers, PZ, "The Sheldrake Phenomenon," *Pharyngula.* Available from: http://scienceblogs.com/pharyngula/2008/06/08/the-sheldrake-phenomenon/ (Accessed Jan 2 2014)
22. Sheldrake, Rupert, *Dogs That Know When Their Owners Are Coming Home: And Other Unexplained Powers of Animals,* New York: Three Rivers Press, 1999, pp 54-63
23. Forgacs and Newman, p 125

Chapter 7

Deus ex Machina

Eyes open to arid plain, high wispy clouds suspended above a horizon sketched out by rocky ridge formations.

The Dawn of Man.

As the ape people carry out their daily tasks, unwanted guests from a neighboring tribe show up to oust them from their watering hole. They stand firm. Not your regular pushover hippie ape people.

At night the almost human eyes are locked in place, spell-bound by an unseen prowling predator.

Come morning the proto-hominids awaken to the spectacle of a black rectangular slab parked in the middle of their outdoor living room. They flip out. The more courageous among them venture close enough to touch it, just for a moment, and scurry off. We see it as if crouched in front of it, looking up at its metal-smooth edge under a crescent moon.

A solitary male toys with a bone from a sun-bleached carcass. You get the feeling the monolith is watching him, perhaps even directing his thoughts, especially when he raises the bone and brings it down like a hammer against the skull of the skeleton, pure id gratification lighting up his semi-human face.

The neighbors are back, this time more insistent. Armed with his bone, our hero strikes the enemy alpha in the head, driving him to the ground for good. Exalted at his status as the new big ape of the savanna, he hurls his weapon skyward, which flips end-to-end until it's no longer a fossil but a nuclear-armed satellite.[1]

In 1968, when Stanley Kubrick's *2001: A Space Odyssey* debuted, the emergence of our species was commonly assumed to have ushered in a great leap in technical mastery. But even then the evidence suggested otherwise, and by now it's clear nothing of the kind occurred for eons both preceding and following the arrival of Homo sapiens on the scene about 2000 centuries ago.[2] Abstract reasoning, symbolized in the film by the perfectly geometrical and colorless monolith, left no mark on our early development. The great cultural and tool-making revolutions took place mostly within the last 70,000 years and didn't really pick up steam until much later. The use of weaponry in tribal conflict turns out to be a recent innovation, perhaps no more than 12,000 years old.[3] Militarized and propertied patriarchy, the real "dawn of man," didn't exist for 95% of the history of the human race.

Where Kubrick was dead-on was his depiction of science as the modern religion. Space flight isn't just a matter of rockets and life-support systems. It's about traveling to heaven, to see it up close and ratify its existence for those stranded below. According to David F Noble, author of *The Religion of Technology,* "lift-off" is code for the Christian concept of Ascension.[4] Recognizing this association during the preparatory moon shots of the 60s, Kubrick went so far as to drop a cosmic cross into his picture, with the momentously aligned moons of Jupiter providing the vertical backdrop for an intergalactic doorway tumbling on its side through space. Kubrick set his film in 2001, the millennial year, because the millennium is when Christ is supposed to return from heaven to take dominion over the earth. In *2001*, this is represented by the return of the astronaut after he's lived out his mortal existence "beyond the infinite." In this ultra-modern age, instead of angels we have space aliens to show us the way to our predestined transcendence.

Christian doctrine has long regarded the mind as the part of us that reflects God, while the body is the corrupting influence that conceals our link to the divine. Following this logic, algebraist George Boole concluded that a mathematical analysis of human thought would provide a blueprint for the mind of

God. Though we don't think mathematically, our reasoning can be expressed in terms of algorithms, sequences of commands that form the basis of computer programs. The reduction of human thought to abstract logic cleared the way for its mechanical simulation, as if the immortal mind, the divine aspect of our being, had been placed in a metal box and switched on.

The first time we see Hal, he's a single red eye reflecting the spinning module whose steady rotation approximates gravity for its passengers. If HAL 9000 has a body, it's the spaceship Discovery, essentially a spine joining a brain to a propulsive tail. His authority is so complete that the two crewmembers seem to have nothing to do during the ride. In his narcotizing purr, Hal explains to an interviewer from "BBC 12" why there's nothing to fear from this eerily placid arrangement.

The 9000 series is the most reliable computer ever made. No 9000 computer has ever made a mistake or distorted information.

The expedition is intended to find out why a perfectly geometrical black monolith, recently dug up on the Moon, emitted a high-frequency pulse directed at Jupiter only moments after the latest big ape, an American bureaucrat, touched it with his spacesuit glove. None of the crew know the purpose of the mission. Only the HAL 9000 computer has been clued in about the unfathomable alien intelligence that has inexplicably called upon the human race for a rendezvous over the fifth planet.

After hinting at the true nature of the mission to Commander David Bowman, Hal makes his first mistake. He claims an antenna is about to malfunction, but when the device is retrieved, nothing's wrong with it. He compounds this error by insisting that the antenna is indeed broken and will reveal its hidden defect if placed back in operation. Bowman and his copilot agree that if the antenna doesn't fail, they'll have to disconnect their computer and operate manually the rest of the way.

We sympathize with Hal. After all, we've got a lot in common. As computer science pioneers Allen Newell and Herbert Simon put it, "the programmed computer and the human problem-solver are both species belonging to the genus information processing system."[5] Of the two "species," the superior is clearly the mathematically rationalized immortal brain. How insulting it must be for Hal that these human beings, made in the image of *apes*, think they're fit to carry out this exalted mission. So when he decides to systematically exterminate the passengers and crew of the Discovery, Hal is being *perfectly rational.*

His real error is that he can't admit to error. When Bowman outmaneuvers him and begins shutting him down, it's sad, even wrenching, yet also satisfying, something we've all secretly wished for: the ability to simply unplug that unhinged egomaniac who has to be in control of everything.

When the time comes to journey through the cosmic doorway, it's the human, not the metal box, that goes through. Hal thinks he has all the answers, the keys to the kingdom, but it's all in his "mind."

I am putting myself to the fullest possible use, which is all, I think, that any conscious entity can ever hope to do.

While there's no reason human emotions can't be mimicked electronically, derangement is something that occurs exclusively to conscious minds. It's significant that scientifically literate audiences can watch *2001* and accept as perfectly plausible the notion that a computer could become mentally alive.

As with any machine, every event in a computer is an effect in a causal chain, and nowhere in this sequence is there self-existence. Unlike you or me or any living thing, a computer doesn't have its own time but merely participates in the general background time of the universe. It has no ongoing presence, no living history it can draw from to make intelligent choices.

Science has devolved into a vast computer made of people, its program to establish authority over every conceivable facet of existence. Whatever is out there, the science machine will locate, isolate, analyze and exploit, disregarding effects on nature or the human race, both of which must submit to its commands. Even the conscience of the researcher has no bearing on the machinery of the system.

We know who programmed Hal. *My instructor was Mr. Langley, and he taught me to sing a song*. But where did science get its mechanistic imperative?

As an ideal, science is the search for truth. But actually existing science, according to sociologist Luciano Pellicani, is an offspring of capitalism, which Pellicani defines as the subordination of material sustenance to the logic of the market. In the newly-liberated city states of the twelfth and thirteenth centuries, the emergence of trade as the basis of economics had the effect of quantifying goods and services in terms of money. Every decision made by the Venetian *mercator* was predicated on the costs and benefits of its implementation. "It was capitalism that made European life rational, imposing a mental habit in which quantitative thought, rigorous computation of resources, respect for the impersonal laws of [market] interplay, had an ever increasing role." The market-based system is a "mechanism for rationalization," which generates a "universe impregnated with mathematics."[6]

Science inherits unexamined the incoherent philosophy underlying capitalism, which combines materialism with abstract market principles that regulate economic transactions. Simultaneously hardcore materialist and wide-eyed transcendentalist, the scientist never wonders how the tangible universe coexists with the timeless mathematical dicta by which all energetic transactions are said to take place.

Like the scientific ideology it spawned, capitalism is reductionist, atomizing society into workers who compete to earn a living and companies that compete to earn a profit. Of foremost importance is servicing the moneyed class and its corporate leviathans, though even investors and executives are sometimes sacrificed for the needs of the corporation, an event played out metaphorically in *Space Odyssey*. HAL 9000 is made in the image of a corporation, not only relentlessly pursuing rational self-interest but immortal. If the imperfect humans infesting the machine run it into the ground, the corporation simply enters bankruptcy and re-emerges with new ownership as if nothing happened. So long as the Discovery stays on course, what's the big deal if its crew and passengers have been killed off?

The very notion of *life* burns out the logic circuits of the science machine. The guiding principle of biology is that life is a primitive notion to be replaced with a more sophisticated mechanistic framework. An organism is a peculiarly inefficient and ultimately self-destructive apparatus.

Originally this kind of made sense. When Descartes postulated the mechanistic theory, pretty much everyone in the West believed in God. If organisms are machines, the mechanic must be the Almighty. It was God who crafted each species and then stepped back as the little gizmos began reproducing. Even before Darwin, however, scientists began to realize that life-forms *evolve* into being. Theorists began focusing on what kind of natural forces could have shaped these odd contraptions. How does a machine manufacture itself?

If the body is strictly mechanical, we ought to be able to trace every physiological event back to its first cause. Ultimately, we should understand how the properties of particles and atoms and molecules account for the functions of cells and organs. Shortly after Crick and Watson's discovery of the structure and function of DNA, Walter Elsasser set out on just such an investigation.

It wasn't long before he hit a brick wall. Considering the variety of atoms in a single cell, the number of bonds that can form between them and the reversible enzyme-driven reactions continually shuffling and re-shuffling them, the number of possible molecular events at any given moment and location in a cell is incalculable. Mathematical analysis, the linchpin of physics, is short-circuited in the labyrinthine interiors of cell, organ and organism. "There is no series of actual experiments," observed Elsasser, "and not even a set of suitably realistic thought-experiments such that it would be possible to demonstrate the way in which all the properties of an organism ... can be reduced to consequences of molecular structure and dynamics, the latter being controlled by the laws of physics." The ultimate origin of physico-chemical processes in the body is "buried in unfathomable complexity."[7]

When Elsasser told biologists to stop trying to ape physics and find a methodology appropriate to living systems, he was politely ignored. Life is messy. Hal prefers order.

Where are the instructions that enable a chain of amino acids to twist and bend into a functional protein? What keeps the cell's myriad self-catalyzing chemical feedback loops from going off the rails? For that matter, why don't the tissues and organs of our bodies simply disintegrate? Why don't we all just fall down dead?

Biology has no answer. Despite proclaiming that they've isolated the double-stranded self-replicating secret of life, molecular biologists can't explain how an organism develops and maintains its characteristic forms or why it even lives at all. Yet they go right on smiling to themselves in their perfect genius as they learn to make pigs aerodynamic and potatoes introspective.

We are all, by any practical definition of the words, foolproof and incapable of error.

It's not as if genetic engineering in any way proves that genes comprise a kind of blueprint for bodily structures. Imagine if audio technicians went around claiming that the music on your radio originates from within the radio itself. Let's say 97.3 FM is a country station, while 98.5 is oldies. Somehow "97.3" contains within it the information required to produce all the music on the country station, while "98.5" is a powerful and mysterious "code" accounting for the rock and roll songs.

Contemplating the virtually infinite choices confronting the organism, Elsasser realized it makes decisions based on past activity. From sub-cellular functions on up, each system works as a whole. Every level — macromolecule, organelle, cell, tissue, organ and systems such as immune, endocrine and cardiovascular — is embedded with the "holistic memory" that helps it navigate endless fields of possible actions.

This anarchistic system of operation doesn't sit well with the science machine, which is offended by a type of memory that doesn't involve computer-like storage of data. Even worse, holistic memory defies the principle that all things are governed by timeless quantitative laws. By contrast, memory originates within time and can evolve over time, sort of like the living things themselves.

It makes sense, but it doesn't compute.

In the late nineteenth century, the introduction of barbed wire fences caused countless injuries among farm animals. According to naturalist Roy Bedichek, at one time virtually every horse in Texas bore scars from encounters with barbed wire. But within 50 years the equine mind was completely adapted to this new environmental condition. Even young horses never exposed to barbed wire know better than to tangle with it. As Sheldrake points out, the effect has become biologically ingrained.[8] Yet the likelihood of a random genetic mutation producing an adaptive instinct in such a short time-span is virtually nil. The memory is holistic, not molecular, relying on the experience of previous horses rather than genetic hocus-pocus.

Darwin makes a similar point:

> Every one knows that hard work thickens the epidermis on the hands; and when we hear that with infants, long before birth, the epidermis is thicker on the palms and soles of the feet than on any other part of the body ... we are naturally inclined to attribute this to the inherited effects of long-continued use or pressure.[9]

It's bad enough that neo-Darwinism rejects its namesake's key observations on the inherited effects of use and disuse within a given species. Far worse is the assumption that wholesale reorganization of body-plan, as with the emergence of fish, reptiles and mammals, is generated through blind alterations in genes. Theorist Jos Verhulst points out that the advent of Homo sapiens required several coordinated developments, including not only the enlargement of the brain but modifications in limbs and larynx.[10] The probability of such a multifaceted transformation being driven by chance mutations is vanishingly small.

Right about now we're beginning to wonder if Hal has a screw loose. Without any basis in observed fact, and in perfect opposition to our senses and intelligence, the science machine tells us life is a series of accidents, a slip-up on the way to mathematical abstraction and its cybernetic animation. Regardless of how absurd its conclusions, the mechanistic mindset cannot be questioned. If we disagree, the mistake must be ours.

This sort of thing has cropped up before, and it has always been due to human error.

Like Elsasser, Sheldrake seeks to explain life on its own terms rather than substituting it with organic machinery and explaining that instead. For the science machine, human memory is akin to stored information in a computer. While the genome preserves the body's memory, the brain stores the mind's. Sheldrake refuses to divide up the matter so neatly. It's the body as a whole, not the brain, that remembers how to ride a bicycle, not because the information is somehow stored in our muscles but because the current effort puts body-mind into resonance with previous rides.

The question neuroscientists like to ask themselves is whether the brain is sufficiently complex to encode the information from which the mind is made. The answer is of course a foregone conclusion. They point out that the brain consists of a hundred billion neurons, on the same scale as the number of stars in the

Milky Way, each neuron interacting with up to a thousand others. This certainly seems encouraging. But if they asked themselves another question, whether this even makes any sense at all, they'd get a very different answer. If the brain really encodes information, we'd need a translator to access our own thoughts. Enter the homunculus, the "little man" who mediates between us and our brains. Despite Descartes' helpful suggestion to look in the pineal gland, no such entity has ever been found.

The only context in which a brain is informative is when it's being monitored by a scanning device. Since this is precisely how neuroscientists approach the brain, from the outside, it's no wonder they think of neural activity as a code to be deciphered. Every neuroscientist is a would-be homunculus — only in someone else's skull.

Brain is mind repeated. We make choices that become habits, and with enough repetition those habits become neurologically ingrained, like depression turning clinical as it sinks into synapses. Once a sufficient number of people have established a given habit, it becomes genetically ingrained, shaping the brain even during embryonic development.

The standard view of the mind has all the mechanisms of mentality but none of the substance. It's got stored information but no memory, perception but nothing that perceives, emotion but nothing that feels. Ultimately the problem goes back to Einsteinian physics. We've got existence but no self-existence, time but no presence, reams of relativities but nothing intrinsic.

Biologists typically fail to follow their materialistic reasoning to its inescapable conclusion: since the self is intangible, it must not exist at all. Only the rare individual is willing to take that final step into oblivion. "What we are, really," says philosopher Daniel Dennett, "is the information contained in our brains." For reasons that remain unclear, the brain hallucinates itself as a "mind." Foremost among the myths of this mind is "the idea of a distinguishable, separate thing called, by some, the self."[11]

In other words, you don't exist; your brain just thinks you do. The mind is no more real than its own images. "What you are is an abstraction ... an abstract object which is definable in terms of a certain set of information."

Of course, some people don't like being told they're "organic robots created by a research and development process called natural selection." They feel this drains life of meaning. But Dennett has little patience for silly ideas about an essential core reality of what we are that provides meaning and direction to our existence. This is, he argues, "an insane idea."

The professor doesn't want any naive concepts to get in the way of his favorite theory. If we are indeed information, you see, we'd be immortal if only we knew how to preserve it, "and there are lots of ways of preserving information." Perhaps when Dennett is downloaded into a computer equipped with an "emotional word processor," he'll be the greatest philosopher that ever "lived." But before dropping to our knees in praise of DAN 9000, let's consider for a moment the nature of computation.

Computers supposedly contain oodles of information. But when we examine the circuitry of one, we don't find any. What we find instead is silicon and wire and electronic impulses. The contents of a computer are no different fundamentally from those of any other physical object. There's nothing general or abstract about a computer chip or relay switch or motherboard. Each object inside the computer is quite specific and definite. Engineers can complicate the wiring as much as they want, but it's still just a bunch of wires. We associate information with the computer, not because there's any such thing located inside it, but because information appears in our minds as we operate it.

Limited to blindly following physical principles, computers must carry out each task in the most elaborately obvious way possible. If birds in a flock had to continually solve, in real-time, the equations necessary to judge their directions and velocities against those of their neighbors, their brains would blot out the sky.

As far as we know, brains aren't computers, and mentality doesn't adhere to circuitry. Mechanists see information in genes and brains and computers but it's all in their minds.

Dennett seems to think our capacity for imagination means we ourselves must be imaginary. It makes you wonder: if we can deny our own existence, is there any limit to our capacity for denial? We're not liquidating ecosystems, human and otherwise, or tweaking genes without the slightest idea of what we're doing. We're not poisoning our air and minds, stripping the ground of soil and water and rendering ourselves dependent on ever more complex and expensive technological solutions to the crises of our own making. Nope, we're just promoting free trade! It's all just information, you see, and there are lots of ways of manipulating information.

Yet even self-negating theorist Dennett has nothing on nuclear physicist Edward Teller, a.k.a. Dr. Strangelove. For it was Teller who showed us the way to actualizing the suicide formerly confined to philosophy.

Father of the hydrogen bomb, a prime mover in the massive US stockpiling of megaton fusion devices, Teller triggered the Soviet buildup of thousands of its own warheads aimed at the US. Dutifully pursuing the flawlessly rational lunacy of the science machine to its natural endpoint, he immersed himself completely and irrevocably in the appalling coldness of pure calculation.

For there's no such thing as life, no such thing as humanness or meaning in this world. *There's only me*. The only thing I understand is that I exist, my sole concern maximizing personal security and power. I'm the one crouched over the skeleton, toying with the bone, looking it over, weighing it. I'm the one smashing the skull and hurling my club into the sky, screeching ecstatically in my supremacy. I must wield the mightiest weapon in the world, the mightiest club there's ever been.

Teller hated anything that might obstruct the proliferation of his beloved offspring. To sell the public on thermonuclear warheads, he conceived of numerous civilian applications for them, such as mining harbors, modifying troublesome weather patterns and exposing urban populations to health-giving low level radioactive waste. He convinced Ronald Reagan that Soviet missiles could be knocked out of the sky with our own nuclear explosives.[12]

Though claiming to act for the betterment of mankind, Teller was driven by craving for prestige and security and power. Like HAL 9000 he sought godhood, and he didn't care if he had to kill us all to get it.

Science started from the idea that we'd just go ahead and be sensible for a change. We'd forget all those superstitions and allow the subject matter itself to define our opinions on it. That the science machine has taken this germ of reason and made world killing insanity from it only goes to show how far we've gone astray.

Open the pod bay doors, please, Hal.

1. Ager, Rob, "Kubrick: and beyond the cinema frame, an in-depth analysis of 2001: A Space Odyssey," 2008. Available from: http://www.collativelearning.com/2001%20chapter%208.html (Accessed Jan 2, 2014)
2. Mithen, Stephen, *The Prehistory of the Mind: The Cognitive Origins of Art and Science*, London: Thames and Hudson, 1996, p 116
3. Ehrenreich, Barbara, *Blood Rites: A History of the Passions of War*, New York: Metropolitan Books, 1997, p 117
4. Noble, David F, *The Religion of Technology: The Divinity of Man and the Spirit of Invention*, New York: Knopf, 1997, p 118
5. Newell, Allen and Simon, Herbert, *Human Problem Solving*, Englewood Cliffs NJ: Prentice Hall, 1972, p 870

6. Pellicani, Luciano, *The Genesis of Capitalism and the Origins of Modernity*, New York: Telos, 1994, p 176
7. Elsasser, Walter, *Reflections of a Theory of Organisms*, Baltimore: Johns Hopkins, 1998, p 3
8. Sheldrake, Rupert, *Dogs That Know When Their Owners Are Coming Home: And Other Unexplained Powers of Animals*, New York: Three Rivers Press, 1999, pp 309-310
9. Darwin, Charles, *The Variation of Animals and Plants Under Domestication*, London: Murray, 1875, vol. 2, p 356
10. Lockley, Martin G, "Evolutionary Dynamics of Consciousness," *Journal of Consciousness Studies*, Vol 17, No 9-10 (2010), p 109
11. Kayzer, Wim, *A Glorious Accident* (TV series), Thirteen/ WNET, 1996
12. BondGraham, Darwin and Parish, Will and Robinson, Nicholas "California's Nuclear Nexus," *Z Magazine*, December 2009, p 22

Chapter 8

Autodelusion

In 1950 the journal *Mind* featured a proposal for determining if computers can think.[1] Submitted by code-breaker and computing pioneer Alan Turing, the idea was to pose questions to a pair of test-takers hidden from the examiner, with the answers from both respondents — one human and one machine — displayed on a monitor. If the examiner failed to determine which of the two was cybernetic, the computer was deemed intelligent. Turing predicted that by the end of the twentieth century, computers would be able to fool the average interrogator 30% of the time if questioning was limited to five minutes.[2] Man-made mind was on its way.

In the movie *Blade Runner* a young woman turns out to be a computer-operated "replicant." It takes more than a hundred questions from a detective to expose its identity. Turing would have been proud. So's the manufacturer, who's not the least disturbed by the fact that the replicant doesn't know it's not human. "Commerce is our goal here at Tyrell. More human than human is our motto."

Contrary to mechanistic expectation, distinguishing computers from people has proven so easy that any computer programmed with CAPTCHA can screen out "bots" seeking access to websites. Granted, killer bots from a "deathmatch" videogame have fooled people into thinking they're up against others of their kind, but then no one was firing philosophical bullets at the bots.[3]

Point being that even if a computer somehow passes the Turing test, a convincing imitation of intelligence is still just that — an imitation. To believe the computer is actually smart is to overlook the distinction between image and reality, in other words, to fail the sanity test.

Nonetheless the dream of digital thoughtfulness lives on.

"What happens when machines become more intelligent than humans?"[4]

So wonders philosopher David Chalmers, apparently unaware that intelligence is precisely where thinking ceases to be mechanical. Undeterred by irksome reality-based considerations, Chalmers not only takes the concept of artificial intelligence at face value but believes AI will evolve, much like the actual intelligence it mimics, into something far beyond anything implemented by human engineers. Writing for the *Journal of Consciousness Studies*, Chalmers argues in favor of a scenario proposed in 1965 by statistician IJ Good:

> Let an ultraintelligent machine be defined as a machine that can far surpass all the intellectual activities of any man however clever. Since the design of machines is one of these intellectual activities, an ultraintelligent machine could design even better machines; there would then unquestionably be an "intelligence explosion," and the intelligence of man would be left far behind.[5]

"The key idea," writes Chalmers, is that a machine "more intelligent than humans will be better than us at designing machines."[6] It's a simple matter of time before an electronic brain outsmarts its biological forerunner, uncovering design principles we never thought to look for. Once next-gen AI arrives, it devises even better ways to design and craft computer circuitry, triggering yet more sophisticated AI and so on and so on. The genie has left the bottle.

In keeping with the trend of exponential gains in processing speed over the past few decades, each generation of AI is expected to improve on its design faster than previous generations, reducing intergenerational downtime from years to minutes to microseconds until at last we arrive at the fabled Singularity, a blast of infinite intelligence yielding a new cosmos composed of bits in place of atoms.

Like most prophets of AI apocalypse, Chalmers never defines the central term in his analysis, even contending that the case can be made without invoking a "canonical notion of intelligence."[7] By avoiding any consideration of the issue, he defaults to the mechanistic view equating intelligence to the algorithmic operations of a computer.

An algorithm is a sequence of instructions for reaching a conclusion regardless of the starting point or obstacles along the way. Let's say you're driving a car to the north end of town. If you're going east the onboard computer tells you to take a left. Going west it tells you to take a right. If you can't take a right, it tells you to turn around, go back and then turn left.

Though necessary for intelligence, the ability to execute a well-defined sequence of actions is nowhere near sufficient. Notably absent from computers are the human traits of understanding and imagination.

Understanding is not a programmable task. Consider a primitive cybernetic system such as James Watt's steam engine and its "governor," which utilizes an algorithmic "if-then" process to stabilize energy output. If the steam pressure rises beyond a threshold, the engine slows down, just as it speeds up whenever the pressure falls below a certain level. At no point does the engine reflect on its varying predicaments and act accordingly. By design its corrections are entirely mechanical. A computer is only a more complex version of this. Though humans can interpret the results, the computation itself is no more thoughtful than a hammer hitting a nail.

Because every operation in a computer is binary, calculations involving the number zero require less processing when it's "denormalized" into a pair of positive and negative zeroes.[8] Yet treating zero in these terms makes no sense and could only interfere with human calculations. It's not a problem for computers only because comprehension plays no role in their operations to begin with.

AI is artificial not just in the sense of artifact but *fake*. Without comprehension, without the ability to reflect on what a statement means, computers can only mimic intelligence. Likewise, if human intelligence is wholly computable, we too are fakes. Incapable of true understanding, subject to digitization and downloading into appropriately configured circuitry, we would have no concept of meaning. Indeed the meaning of meaning is that we don't compute.

As Roger Penrose notes, computer science originated as a branch of mathematics. The original idea was to build an "idealized mathematical 'machine.'"[9] Alas David Hilbert's dream of freeing all correct mathematics from contradiction was famously punctured by Gödel's theorem, which demonstrated once and for all that truth itself cannot be formalized mathematically and materialized electronically.[10] Because a complete axiomatic system must contain statements neither provable nor disprovable within the context of that system, someone has to come along and reflect on its meaning. Only by way of conscious examination, says Penrose, can we "leap outside the rigid confinements of any formal system to obtain new mathematical insights."[11]

Among the greatest achievements of mathematics is the complex number, which combines a "real number" with an imaginary counterpart. Denoted by the symbol *i*, this number equals the square root of -1, a paradox since no number multiplied by itself yields a negative. Though illogical, *i* plays a critical role in physics, including quantum mechanics and the analysis of self-organized *complex* systems. By stepping outside the bounds of actuality, mathematicians retrieved a tool for investigating the more perplexing aspects of the world.

A warped mirror image of understanding, imagination turns

out to be essential to the scientific project. Far from a series of if-then statements leading mechanically to the correct conclusion, Einstein's discovery of special relativity was prompted by a fantasy of traveling alongside a wave of light. How can a computer, which knows nothing of "reality," break through the firewall to unreality?

Computer architecture makes no room for creative leaps. Instead of emerging all at once, discoveries must be built up piecemeal, like trying to reach the other end of the universe in a rocket: no matter how fast you go, you'll never get there. Creative intelligence is like a wormhole whose interior is unreal, at least in relation to space, and this property allows the wormhole to pin a piece of local space to a distant cosmic locale.

But Chalmers has an ace up his sleeve. Surely computers would be conscious and capable of creative leaps if we built them to operate just like the human brain. Like Ray Kurzweil, author of *The Singularity Is Near,* he thinks that by reverse engineering the brain, we can derive the design principles of intelligent thought. The so-called neurosynaptic chip is thought to herald the approaching realization of this very goal.[12]

"If a system in our world duplicates not only our outputs but our internal computational structure, then it will duplicate the important internal aspects of mentality too."[13] Though Chalmers concedes "we have no idea how a nonbiological system, such as a silicon computational system, could be conscious ... we also have no idea how a biological system, such as a neural system, could be conscious."[14]

Perhaps we don't know how neural systems could be conscious because they're not. The only thing we know for sure about the brain is that it facilitates the consciousness of the organism as a whole. If brains themselves are conscious, they're not telling anyone.

Central to the fantasy of the Singularity is the proposal that the brain, like a computer, is a machine. "The weight of evidence to date," writes Chalmers, "suggests that the brain is mechanical."[15] Yet the evidence for mechanical operations within the brain in no way suggests that overall brain function is mechanical. Neural networks operate only in the larger context of a cycling self-organized system. Brains resemble chaotic weather systems far more than predictable machines.

To identify conscious properties in the brain, we must decode it, literally learning its language and reading out the contents of a mind like data from a hard drive. So long as we depend on correlating brain activity with subjective reports of mental activity, we have no basis for reducing mind to brain or brain to machine.

One way around this problem is to assume the false dichotomy of materialism vs. dualism, by which the conscious self is either reducible to the brain or autonomous and distinct. Given the undivided wholeness of the organism, the autonomous mind is easily expelled, leaving only gray matter.

But what if "brain" and "consciousness" are only different aspects of the same thing? Rather than belong to the thing itself, its dual aspects arise from our differing perspectives on it, like viewing a coin from the "heads" or "tails" side. Just as mass and energy are two sides of a coin called matter, the mind appears alternately as brain or conscious-self depending on whether the point of view is external to ourselves or internal. The meaning of mind is the singularity of matter and consciousness, the equality of subject with object.

Human instinct, including language, carries a built-in bias for the visual and therefore spatial. The very idea of perspective is spatial. But when we try to pin the mind to a particular location, all we get is nervous tissue. To flip the coin, to find ourselves, we must take the "point of view" of time. Only in the context of past neural activity does current activity constitute thought. Only in the context of memories and goals does the present take on meaning. Consciousness and presence are one and the same;

consciousness just does it consciously.

We are made of time. Life is the seamless translation of past, present and future into memory, sensation and will.

When it comes to a computer, however, there is no other side of the coin, only a spatial set of material-mechanical components. Lacking its own memory-infused awareness, there's no perspective from which the computer resolves into mind. A computer is a coin which, when flipped over, disappears.

"Metaphor" is Greek for *carry beyond*. A metaphor represents something real, such as a computer, with something unreal, like a coin with only one side. Being illogical, metaphors don't compute. This is problematic, especially since metaphor is implicit in all language. Every word, whether spoken, written or gestured, is at once concrete and abstract. Not only imagination but language itself depends on the ability to carry the tangible beyond the boundary of the real, an impossibility for a "one-sided" object.

Whereas humans are present to events, computers are merely part of the events. What we share in common is a propensity for generality. Jos Verhulst equates human nature with retention of the unspecialized condition.[16] At each stage of primate evolution, our ancestors were less specialized than their relatives, whether australopithecines compared to chimpanzees, Homo ergaster compared to Homo habilis or Cro-Magnon compared to Neanderthal. To be human is to keep to the trunk or central leader of the evolutionary tree, refraining from stepping out onto a branch of specialized function. As ethologist Konrad Lorenz put it, the "true man" is defined by "always remaining in a state of development."[17]

The defining feature of the computer is its ability to simulate any mechanical process, to reach into any branch of specialized function without sacrificing its global flexibility. It wasn't called a Universal Turing Machine for nothing. For this reason AI seems at first glance the ideal vehicle of post-human evolution.

But there's a catch: all the operations performable by a computer have to be well defined in advance. Programmers can generate algorithms for what we already know, but the essence of evolution is the emergence of what no one saw coming. Devoid of comprehension or imagination, AI can only elaborate on already known possibilities of development. No matter how rich those possibilities are, successive iterations cannot amount to anything more than a branch of a branch of a branch, ending not with a big bang of intelligence but the most perfectly refined twig in the universe.

Isolated from questions of life and mind, physics is becoming just such a branch, diverting from the central leader of knowledge into increasingly elaborate and unverifiable theories. In the mechanistic mindset of biology, a branch of a branch, evolution is commonly misconstrued as the environmental selection of competing genetic algorithms. Yet novel behaviors, as Penrose points out, require a "judgment-forming" process, which he characterizes as the "hallmark of consciousness." Only when a task has become second nature can it be carried out algorithmically.[18] Whereas conscious judgment is at the cutting edge of evolution, algorithms merely mark the trail left behind.

Given that evolution is a natural process, we ought to consider that nature's logic differs sharply from the version we installed in our minds and later our books and circuit boards. In contrast to logic 2.0, the logic of nature seems paradoxical: order emerges from chaos; muscle increases with use; mind is both chemical and meaningful, etc. In the present context the key paradox is the *self-organized* complex system.

In the Newtonian dream, both subsequent and prior conditions can always be deduced from current conditions. But the real world includes vast systems of molecules whose behavior doesn't yield to the predictive and retrodictive power of Newton's laws of motion. The behavior of a gas involves way too many variables to be computed on the basis of interactions between molecules. Over time a thermodynamic system obscures its origins, erasing the mechanical "memory" characteristic of a Newtonian system.[19]

When conditions are right, however, natural processes can abruptly switch into coherence, like the systematic transferal of heat during convection. Stupendously improbable from a molecular viewpoint, convection requires only a temperature differential between one end of a fluid and the other, and the system spontaneously organizes to restore evenness in temperature.

Ordinary thermodynamic systems destroy the mechanical retrievability of past conditions, but self-organized systems bring the past to bear in a different way. Taylor vortices, for example, which serve to smooth out pressure differentials in rotating fluids, defy prediction on the basis of current conditions. Only when their history is taken into account are they subject to accurate prediction. Like organisms, vortices carry their past along for the ride.[20] Private memory, in turn, is the basis of interiority and self-existence, the flipside of that dynamically ordered cycling system we call the brain.

According to human techno-logic, memory is the storage of data subject to mechanical retrieval. By contrast the seeming illogic of nature's memory is to obscure the past in a cloud of indeterminate molecules out of which emerges a complex system informed by its own history. Our logic is to build systems, while nature's logic is to let them build themselves. Evolution is a function of nature's logic, nature's complexity and nature's memory. The artificial analogue of these traits, as embodied by the computer, is irrelevant to the future of intelligence.

Far from an explosion of infinite brilliance, the Singularity is a chain reaction of delusion. Since nature is fundamentally mechanical, organisms must be machines; since machines have no self-existence, neither do we; since we don't exist intrinsically, intelligence has no basis in our will to reason but follows blindly from the mechanics of a data-manipulating nervous system; if we can duplicate our own mechanical intelligence by manufacturing an electronic device smarter than we are, it will design a machine smarter than itself and so on ad infinitum. To arrive at the Singularity, you have to think like a machine, with one delusion triggering the next in a fully automated sequence.

Aside from the usual cultural means of propagating delusion, nature is at work as well. Repeated enough, a thought or action becomes habit, and this applies to populations as much as individuals. For instance, when the typewriter was invented, the keys were arranged alphabetically, enabling rapid learning but also frequent tangles. The QWERTY keyboard was designed to avoid these jam-ups. Of course, with the switchover to computers there's no reason to continue with this arrangement except that so many people have already learned it. But it's not just cultural but morphic inertia. Sheldrake reports experimental evidence that learning to type on the QWERTY keyboard, despite the obvious disadvantage, is easier than on alphabetized keyboards.[21] QWERTY seems to be carved into the collective unconscious. When we learn to type on the same keyboard used by predecessors, we're sliding down a groove.

According to William Blum, author of the incisive history of US foreign policy, *Killing Hope*, in the 1920s anticommunist propaganda was carefully crafted to protect elite interests from popular uprising. Strangely enough, by the 50s the disseminators were buying into their own lie as anticommunism mutated from "cynical exercise" to "moral imperative."[22] Injections of Red Scares into the body politic were obsolete, as the propaganda was propagating itself. No need to fabricate evidence that communism was an international conspiracy to destroy American freedom when this very belief was insidiously colonizing our minds from within.

So much easier to slide down the groove than think for yourself. The people who organized the war against Vietnam were called "the best and the brightest," which just goes to show that being smart is no barrier to autodelusion. Even long after the fall of the Soviet Union, anticommunism remains implicit in US culture, revealed by persistent smearing of the late Venezuelan president, Hugo Chavez, though his socialism was nothing like the mangled Soviet version. So too, despite the failure of the Human Genome Project to reveal the instructional manual of life, when the topic of genetics comes up we're likely to resonate with those before us who saw DNA as the secret of life. Unlike most of his successors, Einstein was disturbed by the loss of the now. Only over time did the idea of time's unreality become second nature for physicists.

How many hallucinations have mutated into mental parasites, loosed from the minds that hatched them to feed upon the culture at large? Whether you're a teenager mistaking amplified celebrity worship for music or an industrial farmer confusing petrochemical-infused clay with soil, once a lie picks up enough momentum, it's liable to steamroll right over you. Ultimately the spell is cast not by the liar but the lie itself.

Locked away in our Quantopian vault we're forgetting how to be human, how to perceive and think and evolve. With the severing of our umbilical link to Mother Earth and the decline of community bonds and skilled labor and education for its own sake, each successive generation is less grounded and more vulnerable to self-organized systems of delusion. Whereas proponents of infinite AI see it as humankind's savior from deepening environmental and economic crises, in reality the Singularity is a warped mirror image of the end of the world.

1. Turing, A.M., "Computing Machinery and Intelligence," *Mind*, LIX, No. 236, pp 433-60
2. Penrose, Roger, *The Emperor's New Mind*, Oxford: Oxford University Press, 1989, p 10
3. "A new computer game 'bot' acts just like a real person,"

BBC News Technology, September 27, 2012. See http://www.bbc.co.uk/news/technology-19748209

4. Chalmers, David, "The Singularity: A Philosophical Analysis," *Journal of Consciousness Studies*, 17, No. 9-10, 2010, p 7
5. Quoted by Chalmers, p 7
6. *Ibid*, p 7
7. *Ibid*, p 23
8. See http://en.wikipedia.org/wiki/IEEE_754-1985
9. Penrose, p 35
10. *Ibid*, p 102
11. *Ibid*, p 110
12. Sims, Kelly, "IBM Unveils Cognitive Computing Chips," *IBM News room*. Available from:http://www-03.ibm.com/press/us/en/pressrelease/35251.wss (Accessed Jan 2 2014)
13. Chalmers, p 15
14. *Ibid*, p 44
15. *Ibid*, p 16
16. Lockley, M.G., "Evolutionary Dynamics of Consciousness," *Journal of Consciousness Studies*, 17, No. 9-10, 2010, pp 102, 106
17. Quoted in Gould, Stephen Jay, *The Panda's Thumb: More Reflections in Natural History*, New York: Norton, 1980, p 107
18. Penrose, pp 412-414
19. Schneider, Eric D, and Dorion Sagan, *Into the Cool: Energy Flow, Thermodynamics and Life*, Chicago: University of Chicago Press, 2005, p 72
20. *Ibid*, pp. 129-30
21. Sheldrake, Rupert, *The Presence of the Past*, New York: Times Books, 1988, p 195
22. Blum, William, *Killing Hope*, London: Zed Books, 2003, p 11

Chapter 9

Animals and Monsters

The man who never alters his opinion is like standing water, & breeds reptiles of the mind.
— William Blake

Only in the context of civilization is the concept of "whore" even meaningful, much less the predicate of nature. In the modern mindset, now penetrating deeply into the world's many societies, nature is to be drained of every last drop of usefulness and discarded without thought by jealous consumer gods. By this view nature is "other," the same other that infants experience as mother, that which feeds and sacrifices, like the delectable flesh of grazing animals or the labor of the hopelessly poor.

Elaborating on a theme already well abstracted into the general civilized mind, Descartes alleged that animals have no soul or internal motivation or passion or intelligence. All their movements and expressions depend entirely on the "disposition of the organs, no more nor less than do the movements of a clock or other automaton, from the arrangement of its counterweights and wheels."[1]

Not quite. Turns out the power of abstract reasoning turns up everywhere we look in the animal kingdom. French researcher Martin Giurfa sought signs of intelligence in bees and found it by exposing them to a series of Y-shaped mazes, each of which displayed a symbol at its opening. This symbol then reappeared at a branching point, while the other branch displayed a different mark. After passing through various mazes, the bees surmised the honey would always be on the branch bearing the same symbol as the opening. If Giurfa marked the opening blue, bees followed the branch marked blue. If he marked the opening with vertical lines, they knew to avoid the branch marked with horizontal lines.[2]

At least bees have brains. Yet even plants, according to Jeremy Narby, author of *Intelligence in Nature*, rationally respond to variations in "light, water, temperature, chemicals, vibrations, gravity, and sounds ... by changing the way they grow. They can assimilate information and respond on the whole-plant level."[3] As Darwin pointed out, plants readily violate their habitual period of flowering in the course of adjusting to new climates.[4]

Is this so surprising? After all, what is a nervous system if not a flowering bulb atop a stem rooted in flesh and propped up on a stick?

The history of bigotry, if not against the vegetable-other then at least against the animal-other, has triggered an equal and opposite reaction. At last we're embracing our furry friends. Time and again we're told that people are animals, and we nod in agreement lest we be thrown in with creationists, who can't accept that our australopithecine ancestors evolved alongside apes from deep primate stock.

But evolution comes with some big jumps. There's a reason we call bacteria "life" but not cycling self-catalyzing chemical reactions. Sometimes a line is crossed, and a new kind of system presents itself that can't be classified with those that preceded it. One thing nonhuman organisms don't do, whether amoeba, plant, insect or mammal, is mirror the world in symbol and then play with those symbols as if manipulating the world itself.

The instinctive response to invasive ants is to stomp on them. A slightly more clever approach is to poison them. But only a person would think to poison them without immediately killing them so they can return to their nest and spread the toxin among their extended family. We're able to think of this because we can model the situation in mind and then "see" it in a way animals can't.

Parrots can comprehend numbers up to at least six, but only

people would think to invent -1 or an impossible number which, when multiplied by itself, equals -1. Animals inhabit land, air and sea, but to find *i* you need to reflect on mind itself and take up residence in imagination.

Like all living things, each of us explicates our species. Every individual is literally the fleshing out of the order implicit to its kind. Humans take it a step further. We manifest not only our species in body but our culture in mind. We run on parallel tracks, one through the senses and the other through inner reflection. The shared invisible culture we generate through conversation and play is totally unknown to animals, even our closest companions. As we negotiate this galaxy made entirely from shades of meaning, our pets think we're just hooting and whistling and barking.

Human consciousness is the animal mind reflected back on itself, generating an enclosed "space," a conceptual environment where ideas and attitudes take root and evolve. This is not to deny occasional reflectivity among animals. That apes routinely attempt to influence the perceptions of their peers may indicate awareness of each other's minds.[5] But even chimpanzees and bonobos, able to learn the meanings of many symbols, never make use of them to express personal insights.[6]

If apes dipped their toe in the reflective pool, the human race dived in. Like all animals, apes exist as bodies first and minds second. With us it's the other way around. We are mental creatures that happen have bodies.

From the universal self-existence of flowing presence to the local self-existence of life on Earth, humans have added a third stage: the mental self-existence of reflective consciousness. A radically new thing in the cosmos, humankind is understandably proud.

That we've stepped outside the circle of nature is indicated by the very idea of it, which has no possible meaning to an animal. This is not to say we're divorced from our animal nature, only that we're no longer identical to it. We've sailed off into a twilight zone, unable to abandon nature altogether yet no longer truly part of it either.

The transformation from a body that makes use of a mind to a mind that makes use of a body is signified by the end of brain expansion. Having reproduced the world in thought, we can scrutinize beliefs and generate insights with astonishing ease, a capacity that left our human forerunners with far more cerebral processing power than they needed. The animal mind was built to work in the dark, but we've got a light on in there. We sense thoughts and feelings like animals sense food and predators. When it comes to figuring out complex problems, whether social, natural or technological, we cut right to the chase. A conscious insight is worth a thousand unconscious calculations. In light of Darwin's principle of disuse, this might explain why our brain is slightly smaller than that of our cousin, the Neanderthal, who apparently never crossed that line into a self-contained universe of mind.[7]

A prime example of *neoteny*, our species evolved from the childhood form of our predecessor.[8] We start out like archaic Homo sapiens children, but we never progress to the adult version. The offspring of a youth revolt, we keep our small faces and curious eyes for life. We also retain the capacity for imaginative play, the products of which take on momentum all their own. From nothing more than random humming, a melody emerges and takes command.

Aside from our immediate and continual experience, the clearest indication of mental self-existence is the habit of projecting human traits all around us. Storm clouds are mommy and daddy coming back to castigate us. Flamingos are elegant. Cranes do a little dance before mating. Even rivers and trees and rocks are really just like us, each with its own purpose. The river wants to go downhill. The tree wants to reach into the sky. The rock is content to sit quietly. By looking at nature from its imagined points of view, our predecessors gained unexpected practical insights. Paleontologist Juan Arsuaga speculates that this tendency enabled our forebears to outcompete the Neanderthal despite physical disadvantages. Though fully adapted to Europe, the Neanderthal never adapted to imagination and never reaped the resulting cognitive harvest.[9]

In an age of mass production of uniform artifacts, the idea that everything is alive no longer has much pull. Yet anthropomorphic projection persists. Since only we build them, machines serve as a kind of substitute for the human form. To see the logic of the machine at work in organisms, as Descartes did, is to project our technological "soul" onto nature.

No one is born human; each of us comes into this world with the body-mind of an animal. Far from a product of nature, a person has to be cultivated like crops from earth. Not until well into the first year of life does the snake swallow its tail, so to speak, and the self-contained mental organism known as a human being snaps into place.

Given the difficulties of childrearing, parents have been known to abandon their offspring, sometimes to an orphanage and sometimes a garbage bin. In times past, unwanted babies were often left in the wilds and occasionally survived through assistance from animals like dogs or monkeys. The best documented case is that of Victor, who was captured near the French village of Aveyron in 1800 after multiple sightings in the woods. No matter how much care and instruction he received, however, Victor couldn't quite master the basics of being human, including grammatical language and social interaction. Victor inhabited the world but not the repeat-world of culture conjured up from consciousness.[10]

In his lively account of the evolution of thought, *How Homo Became Sapiens*, Peter Gärdenfors keeps telling us we're animals even as he keeps showing us we're not. Fundamental to the human-animal divide is our relationship to time. Whereas animals are merely carried along on the current, humans know the depths all too well. We're not just influenced by past behaviors but capable of intentionally recalling prior events. Whereas animals can devise and carry out plans to meet immediate needs, the rare case of planning for future needs, like squirreling away acorns for the winter, is purely instinctive. No creature *gets* time like we do.[11]

But animals retain a depth of wisdom we humans sloughed off long ago. They never abandoned nature, never stepped out of their lifeways into something wholly alien, never refashioned ecosystems into living rooms. In short, they never took leave of their senses. A meadowlark keeps to her way and lets nature do the rest. Only people are so misguided as to turn our back on actuality and rely on money, a mere abstraction, to care for our needs.

As proton explicates subatomic habit, protein explicates subcellular habit. To this simple arrangement, humans have introduced a twist. In both brain and artifact, we explicate cultural forces bigger than we are, modes of perceiving and acting that evolve their own logic often at odds with reason, sabotaging the generalized intelligence that brought us to the pinnacle of evolution.

Having corrupted the very factor that makes us human, we've lost our way if not our minds. We alone find the fundamental reward of being alive unsatisfying. We alone cultivate an imaginary life centered on a made-up self called the ego, this hungry ghost that never attains fulfillment no matter how many goodies it gets. Unsatisfied with the world that birthed us, we remake environments to be rational, linear, streamlined. Animals seek pleasure and flee pain but never try to change the structure of daily existence to better their odds.

Ancient poet Chuang Tzu, author of *The Way*, encapsulates our dilemma in the parable of two kings and their oddball buddy, "No-Form."

> South Sea King and North Sea King used to go together often to the land of No-Form, who always treated them well. So they consulted and thought up a good turn, a pleasant surprise for No-Form in token of appreciation. "We have seven openings for seeing, hearing, eating and breathing, but No-Form has no openings. Let's make him a few holes." So they put holes in No-Form, one a day for seven days. And when they finished the seventh opening, their friend lay dead.[12]

We're not just reconfiguring outer environments but ourselves, systematically punching holes in the animal within. We seek to mold our nature to match our human idea, to subjugate the beast. Anyone who's ever gone into a rage knows the futility of trying to efface the animal within. Since we can't eliminate our foundation, all we can do is pretend it's not there.

In the act of mirroring the world comes the concept of the good. We dutifully train ourselves to follow what is virtuous, to promote progress and happiness. But the concept of good brings with it the possibility of bad. If we do right because we're acting on an idea of right, before long we get an idea of something else.

The trouble with virtue is how easily it Orwellianizes into justification for killing and stealing and oppressing. Says Chuang Tzu, "Teaching love and duty provides a fitting language with which to prove that robbery is really for the general good. A poor man must swing for stealing a belt buckle, but if a rich man steals a whole country, he is acclaimed statesman of the year."[13] In the framework of "business ethics," corporate CEOs aren't raping the earth and wrecking communities but fulfilling their fiduciary duty to the shareholder.

All of recorded history has taken place during an era of general insanity. Poetry, tragedy, philosophy, music, painting… it's all one long anguished attempt to restore the wholeness of the human mind — so far in vain. Instead of relying on ever more refined systems of laws and punishments to regulate a world gone mad, we must abandon the great conceptual hall of mirrors and face reality. We don't need to be ruled by "superiors" when we can do just fine by simply perceiving and acting according to our nature. That we naturally generate an invisible realm of symbols doesn't mean we have to get lost in it.

What is now called virtual reality is only the latest iteration of a process at work for millennia. We've long sought a world more human-friendly than the one nature gives us, a world that reflects our logic back onto us, where reality is how we program it. In cyberspace the ego feels fully at home. Whatever you think you are, whatever self-image you project for electronic social interaction, that becomes what you "really" are. No need to reflect on regrettable or even shameful actions of the past. With nothing real to remind you of any of it, all that stuff can stay buried, gathering dust in the animal vault at the core of the mind.

In 1492 Christopher Columbus set sail for the Far East, and in his virtual reality that's where he landed. Columbus found natives, the Arawak or Taino, without industry or laws or class division, and he saw his job as civilizing them, putting them to work and making them pray and wear proper clothing. After he wrecked one of his ships, the Arawak felt sorry for him and brought him gifts, including gold adornments. In his virtual reality, this was their way of saying they were good people who only lacked the leadership to bring them to civilization.

But they really weren't "good," at least not in the sense of being self-consciously virtuous. They simply expressed their nature. No need to lead them into the civilized idea of the good when they already lived the real thing. They were fully human, fully present to each other and their surroundings, fulfilled and at peace.

Until, that is, Columbus forced them to dig up enough gold to impress the royalty back home and finance further expeditions.

Never mind that the Arawak had given him treasure accumulated over generations and couldn't possibly gather equivalent amounts on a regular basis for eons to come.

Columbus sold individual Arawak into slavery across the Atlantic, but at least he refrained from interfering in their social order at home. Not until he lost control of his colony to a popular revolt did things really fall apart. The Spaniards who'd accompanied him saw what was to be gained, and they wanted a piece of the action. In their virtual realities, they too were kings like Columbus. All it took was exploiting the Arawak more fully, more rationally, systematically. The result was social disintegration as extended families were divvied up among colonists. Natives who escaped were hunted down with the help of dogs. The Spanish dream was the Arawak nightmare. Refusing to seed the next generation of slaves, in 50 years their population plummeted from hundreds of thousands to a few hundred.[14]

What does it say about the current state of the human race that the original sin of the New World was triggered by popular takeover of Columbus' colony? We are naturally a social species. How did we descend from human race to rat race?

A satiated predator stops hunting. Not so for people under the spell of virtual reality. Since gold represents wealth, and there's no limit to how much wealth can be acquired, the human predator is never filled up and keeps at it until nothing's left to hunt. When reality is replaced with a virtual substitute centered on our sense of entitlement and the image of our own inherent goodness, systematic intelligence becomes a weapon wielded against common sense. The ability to reflect and reason becomes the means by which we maximize the satisfaction of animal craving and rationalize or cover up the costs. It's the worst of both worlds, all the drive of the animal but none of its heart.

In "America the mythical," as the late columnist Alexander Cockburn put it, children enjoy an Easter egg hunt on the White House lawn. Meanwhile in "America the real," a fire in a poultry concentration camp forces the euthanizing of a quarter million egg-laying hens. In America the mythical, Michelle Obama tends to her organic garden, while in America the real her husband appoints Tom Vilsack Secretary of Agriculture so he can defend the corporate agenda of bioengineering crops to produce their own insecticides, dosing animals with hormones to induce excess growth before slaughter and applying ammonia to rotten meat so it can be consumed in school cafeterias.

"Mythical America," writes Cockburn, "is interminably featured in pastoral commercials and cute cartoons, where contentment reigns at Animal Farm."[15] In real America, "animal gulags emit appalling stinks, clouds of methane so toxic that they kill ... Reeking lagoons surround darkened warehouses of animals trapped in metal crates barely larger than their bodies."

In mythical America orca whales happily play in their pools to amuse their audience. In real America these captive creatures damage their teeth gnawing on gates, even occasionally attacking and killing their trainers, such as Dawn Brancheau, who was dragged underwater by her ponytail at SeaWorld Orlando. Since orca attacks like this are completely unknown in the wilds, Cockburn advocates freeing all enslaved sea mammals. "If it's judged by an independent panel that the artificially bred orcas simply couldn't hack it in the wild blue yonder, let them laze around in their pools and toss them an occasional corporate executive, perhaps starting with slave-owner Pete Peterson ... a public pest who richly deserves an orca jaw clamped on his ankle."

The orca follows its natural way by striking back at its captors. The same cannot be said of the capitalist who profits from a creature's confinement and exploitation.

Originally the Beast was conceived as the offspring of Satan and a woman, an inversion of the holy seed planted in mother Mary. The number of the beast, now believed to have been 616, was early Christian code for the Emperor Nero, who carried on the malignant tradition of his predecessors Tiberius and Caligula except that in his virtual reality he was a great artist who fed Christians to lions only because he was so devoted to the happiness of his people.[16] Sure, he dotted the route to his parties with the burning corpses of those who annoyed him, but he made sure to limit the victuals to half a boar so as to avoid senseless waste.[17]

Nero seems to have been stranded at the phallic stage of development. Around age four, lacking a solid base of empathy or reasoning, the child's primary motivation is to express power over objects, including people. The only way to get the child past the phallic stage is through a combination of love and discipline. *Yes, we love you, but no, you don't always get your way*. Nero got lots of indulgence or lots of abuse (and possibly both) but not much in the way of love, and he grew into a monster, no longer animal but not truly human either.

Which brings us back to Pete Peterson, who bought SeaWorld not to free the orcas and dolphins but to load it up with debt so as to pay himself a hefty dividend. One-time Chairman of the Council on Foreign Relations, head of the New York Federal Reserve during the Bush II years, multi-billionaire uber-investor and leader of the "Wall Street deficit hawks," Peterson has long advocated severe reduction in federal social investment, denouncing such spending as "fiscal irresponsibility" even as his own hedge fund was fueling and profiting from the $8 trillion housing bubble. So long as his friends in Washington turned a blind eye to the true crisis, the bubble continued expanding and finally burst, driving millions into poverty and forcing Congress to radically increase deficit spending just to hold the financial system together. Has he apologized for his profoundly misguided agenda? To the contrary he's used the crisis of his own making to intensify demands for obliterating Social Security and Medicare, personally bankrolling a national campaign to cut off struggling Americans from government assistance they've already paid for.[18]

But Peterson's no Nero. In fact he comes across as a real nice guy, very civilized and decent, at least up to the moment he bears his fangs. Peterson represents a refined form of the beast, the vampire, well-mannered and gregarious, carefully orchestrating a dance that winds up with his teeth in your neck. Lacking genuine human fulfillment, he must derive a simulacrum of it by tapping into the lifeblood of others.

Much like a corporation, the vampire is a virtual person, all ego and no soul, an undead walking abstraction devoted only to the accumulation of abstract "good." Every capitalist, whether or not he's personally stuck at a regressive stage of development, plays the social role of the vampire. As Karl Marx put it, "capital is dead labor that vampire-like only lives by sucking living labor."[19] Investors accumulate their riches by extracting profit from the productivity of others. Labor is under the thumb of the concentrated wealth accruing from its past productivity, sort of like being haunted by your own ghost.

As the human race is defined apart from nature, so too the vampire stands outside the circle of humanity, tapping the circulation of wealth instead of contributing to the flow. By siphoning off money, he perpetuates poverty among the many and thus their dependence on him, which he mistakes as a sign of his superiority. Only in his animal unconscious does he know the truth. He fears the light of day because he doesn't want to be exposed for what he is. Placing a mirror in front of him is pointless, as he doesn't reflect.

Upon our birth we are integrated and whole at the level of animal body-mind. With the onset of human mirror-mind, however, the process of becoming whole has to start all over again. Not everyone heals the schism by integrating animal drives into mental life. The result is a grown-up body concealing a baby in emotional development.

Unable to integrate his pain, the regressive individual

offloads it onto a suitable victim, be it human or animal. By imposing pain onto another, he gets a break from his hell of frustration and fear. When the salve wears off, he just victimizes some more, each time repressing any and all reflection on his cruelty, silencing those who might remind him of it. With self-knowledge locked away in the animal vault, he can't recognize himself for what he is. Pretty soon the map bears no relationship to the reality, self-serving crimes effortlessly rationalized into loving acts of kindness.

An animal cannot be devoid of integrity any more than it can go without a body. Only a person, no longer identical to a body, can be crooked or two-faced or gutless or ungrounded or sketchy or shady or simply a mess. Some people are so lacking in personal integrity that a body reflecting their soul would die the moment it hatched.

As the world draws deeper into abstracted commodified existence, the virtual person is increasingly at home and in command while the human person is ever more alienated and marginalized, ultimately to be rendered into latter day Arawak.

The Spanish colonists who killed off the Arawak most likely effaced all sense of their depravity by way of the collective narcissistic pathology known as racism, the same justification at work for centuries as 50 million Africans were captured and shipped across the ocean into slavery. Over time the more Europeans who kidnapped or owned human beings, the more the insanity would have seemed the very picture of common sense. A groove was established in the Euro-American mind as slave owners resonated with their forerunners. Instead of reflecting and absorbing the horror of the slave trade, they let a meme — an autonomous packet of memory — do their thinking for them. As slaves lost their freedom of body, slavers lost their freedom of mind.

The master-slave mindset perpetuated itself as surely as the form of an atom or a microbe or a flower. Once a set of properties takes shape in the implicate order, only intelligent intervention can prevent continued explication.

Nero was at least nominally human. The real monsters aren't so much the regressives among us as our own psychic progeny, these invaders from inner space that bring with them alien imperatives like class war, racism, patriarchy, conformity and inquisition. Like us, autodelusion is a creature of the mind, a member of the kingdom Psyche. Most beliefs that circulate in our societies are as benign as the familiar creatures of air, land and sea, providing the threads from which diverse cultures are woven. Because they're devoid of human perception or sympathy or intelligence, however, the self-perpetuating beliefs to which we give birth can just as easily turn pathological, and these propagate right alongside the rest. Racism, still evident decades after the civil rights movement, is a product of morphic inertia as much as a byproduct of class war.

No animal would ever be caught dead debating whether or not it's an animal. In our bones we know we're set apart. Residents of culture as much as environment — image as much as body — we unknowingly churn the mental compost from which invisible breeds of life advance.

1. Descartes, Rene, *Treatise of man,* Cambridge: Harvard University Press, 1972, p 113
2. Narby, Jeremy, *Intelligence in Nature: An Inquiry into Knowledge,* New York: Tarcher/Penguin, 2005, pp 55-56
3. *Ibid,* pp 83-84
4. Darwin, Charles, *The Origin of Species,* New York: Modern Library, 1993 (1859) p 29
5. Gärdenfors, Peter, *How Homo Became Sapiens: On the Evolution of Thinking,* Oxford: Oxford University Press, 2003, pp 99-103
6. Donald, Merlin, *A Mind So Rare: The Evolution of Human Consciousness,* New York: W. W. Norton & Company, 2001, p

121
7. Arsuaga, Juan Luis, *The Neanderthal's Necklace: In Search of the First Thinkers*, New York: Four Walls Eight Windows, 2002, pp 67-68, 312
8. Mithen, Stephen, *The Prehistory of the Mind: The Cognitive Origins of Art and Science*, New York: Thames and Hudson, 1996, p 63
9. Arsuaga, pp 308-310
10. Itard, Jean Marc Gaspard, *The Wild Boy of Aveyron*, Ann Arbor: University of Michigan, 1962. See also the National Geographic TV production, *Feral Children*. Available from: http://www.youtube.com/watch?v=STn3bpTTU6c&NR=1 (Accessed Jan 3 2014)
11. Gärdenfors, pp 71-76
12. Paraphrased from Merton, Thomas, *The Way of Chuang Tzu*, New York: New Directions, 1965, p 66
13. *Ibid*, p 68
14. Morgan, Edmund S. "A World Too New," *Smithsonian*, October 2009, pp 82-96
15. Cockburn, Alexander, "Feed Pete Peterson to the Whales," *Anderson Valley Advertiser*, March 3, 2010, p 6
16. Grout, James, "Encyclopedia Romana," *University of Chicago*. Available from: http://penelope.uchicago.edu/~grout/encyclopaedia_romana/gladiators/nero.html#anchor5371 (Accessed Jan 3 2014)
17. McDonald, Marianne, "Making the Gods Jealous," *CounterPunch*, Vol 19, No 15-16, September 2012, p 26
18. Brown, Ellen, "IMF-Style Austerity Comes to America" *counterpunch.org*, March 2, 2010. Available from: http://www.counterpunch.org/2010/03/02/imf-style-austerity-comes-to-america/ (Accessed Jan 3 2014)
19. Marx, Karl, *Capital: A Critique of Political Economy, Volume One*, New York: Cosimo, 2007, p 257

Chapter 10

The Experiment Requires That You Continue

Social worker Morris Braverman is doing his best. If only the idiot in the electric chair would correctly answer the incredibly easy questions lobbed at him, Braverman wouldn't have to keep shocking him at ever increasing voltages.

At 75 volts the "learner" grunts. At 150 he demands the experiment come to a stop. At 285 he unleashes a primal scream. Yet the control board goes all the way to 450 volts, the effect of which is indicated with a skull and bones insignia. Meanwhile the creepy guy in the white lab coat refuses to give an inch. In response to Braverman's growing discomfort over his role as "teacher," the scientist tells him he has no choice.

Though described as a "repressed and serious man," Braverman has suddenly come down with a bad case of the giggles. The more agonized the responses from the chair, the more uncontrollable his laughter. He rubs his face to hide it. When that doesn't work, he pounds the table. He truly feels terrible for the guy strapped in, but he can't help it. Too funny. Too damn funny.

He goes all the way to the end of the board, transmitting a lethal dose of raw power three times to the man who by now has fallen silent. Unable to see him, Braverman has no idea if he's just killed someone. The laughter stops. Finally the scientist informs him that no voltage was actually delivered. The guy in the chair was acting. The only real shock is Braverman's.[1]

Billed as an investigation into the effects of punishment on learning, the experiment was actually designed to demonstrate how far people will go in carrying out orders. Braverman is only one of thousands tested this way, 62.5% of whom go to the end of the board just like him.[2]

Stanley Milgram, who designed the experiment, was as shocked by the results as his subjects. He figured only the occasional psychopath would continue to the final button. Since there's no way almost two-thirds of us are diagnosable with antisocial personality disorder, something else has to account for the findings. When Milgram left it up to the subjects themselves to determine what level of shock they would administer in response to wrong answers, virtually everyone delivered the lowest voltages. In one of the more ingenious variations on the experimental design, instead of an authority figure, Milgram placed a seemingly ordinary guy in the lab who'd get the bright idea of administering shocks at ever increasing levels for each wrong answer. Time and again the actual subject of the experiment heroically put a stop to it.

Milgram concludes that the willingness to kill "cannot be explained by autonomously generated aggression but ... the transformation of behavior that comes about through obedience to orders."[3] A clear majority of people will kill a stranger on command of authority so long as they can't see the victim and the method of dispatch is something easy like pushing a button. As a recent French replication of the experiment demonstrated, peer pressure boosts compliance to 80%.[4]

Mrs. Rosenblum, a middle aged housewife, is the very picture of kindness. She volunteers to help "juvenile delinquents." She's a scout mother. She goes to PTA meetings. She speaks often of the importance of love and healing. And she hated – just *hated* – making that poor man scream. It broke her heart to have to continue shocking him even after he went mute, ultimately pressing the skull and bones button three times. Being a nice person is apparently no barrier to going to the end of the board.[5]

The problem, according to Milgram, is "the capacity for man to abandon his humanity ... as he merges his unique personality into larger institutional structures."[6] In the context of malevolent authority, admirable traits such as loyalty, discipline and, yes, even good manners become agents of destruction. "Do I go right to the end, sir?" asks another nice person. "I hope there's nothing wrong with him in there."[7]

Whenever subjects falter at their task, the guy in the white lab coat gives them prompts such as "Please go on" or "You have no other choice" or, most ominously of all, "The experiment requires that you continue," giving the impression that the final authority is not a human being but science itself. As Milgram observes, in the mind of the subject, the experiment takes on "impersonal momentum of its own," as if no one, not even the psychologists who designed it, can stop it now.[8]

Conscience is nullified in the context of authority. We turn inside-out, the basis of our behavior no longer reason from within but orders from above. In the inverted state, we'll seize on any justification for our actions. *It's not my responsibility... everyone else is doing it... even as I do wrong, in my heart I'm better than that*. To act one way while privately thinking another way is called dissociation. Inverts are experts at dissociation.

Years ago, fresh out of college, I got a job at a warehouse moving boxes from a conveyor belt to a pallet. After a few days of this, the shift ended early and we were excused. But when I got to the exit of the building, about 25 workers were standing in a semi-circle staring silently at the exit, as if the door cast a force-field keeping them back. When I asked why no one was leaving, a supervisor stepped out of a nearby office and demanded my name. After telling him, I remarked to one of the other workers that the supervisor didn't answer my question. The kid couldn't even make eye contact much less respond. I was already fired.

Several years later I found work at a "home service" company and eventually got promoted to complaints manager. From San Diego to the Bay Area we fixed plumbing, wiring, heating and AC. It seemed only natural that such a big company would get a lot of complaints. My job was to prevent those complaints from escalating to the Contractors State License Board, which could take away our license to operate. In the course of my work, I discovered that some of our technicians were systematically overcharging customers, even performing unnecessary work. But this time I kept my mouth shut. I did the best I could to divert complaints and protect the company even though it was ripping people off. Eventually the CSLB launched its "accusation" and I was laid off. No longer in the daily presence of my boss, my moral sense returned and I wound up testifying against the company in court.

Milgram ignores the danger of obedience in the private workplace, focusing instead on government bureaucracy and the war against Vietnam. He devotes several pages to the My Lai massacre but says nothing about multinational corporations. Perhaps this is because he was writing at the tail end of the post-FDR regulatory era when the capitalist bug seemed to be contained as effectively as smallpox.

Today the top story is the corporate push to privatize every aspect of human life and to cash in on the earth's bounty before it's all gone. Millions of corporate employees are suspending their conscience five days a week in the service of a CEO. Meanwhile the boss himself has to turn a profit for shareholders if he expects to keep his job. Big government works to maintain the privileged position of big business, which in turn contributes big bucks to the major political parties. Obedience to the authority of their party keeps ordinary people, whether Democrat or Republican, voting against their own interests time and again. Whoever wins the White House, the ultimate authority is capitalism itself, that ongoing experiment in the upward distribution of wealth.

With the global imposition of neoliberal "free trade," farmers throughout the underdeveloped world can't compete with subsidized food imported from rich countries like the US. Unable to feed their families because they have to grow export crops like cotton and coffee, farmers face starvation when their commodities drop in price. So desperate is the situation that ingestion of pesticides is now the leading cause of rural suicides in China and Sri Lanka. "Around the world," writes Raj Patel, "farmers and farmworkers are dying, with the connivance of elected officials, and at the whim of the free market." As Patel points out, we don't want to cause suffering with "every mouthful of food," but that's exactly what we do.[9] As long as we keep buying into the corporate food system — always on the lookout for the lowest price — we're killing complete strangers on the command of the almighty dollar.

Milgram claims our predisposition to obey is a product of evolution. The tribe that follows its leaders prevails over the one where everyone does their own thing.[10] This assumption, however, has no basis in anthropological findings. For nearly 200,000 years our ancestors did quite well in egalitarian tribes, migrating across the earth and adjusting to radically different climates without advanced technology or long working hours. Slavery was no more necessary than royalty. No need for cops and judges and prisons when tribes could make use of low-tech responses to the chronically obnoxious like expulsion or, worse, the silent treatment. Instead of striving to concentrate wealth and power, people pursued collective interest. Loyalty to that interest didn't necessarily translate into blind obedience to leaders.

The great inversion began more than 10,000 years ago with the emergence of war along with class divide driven by a warrior elite. Agriculture soon followed this tectonic shift. While not generally necessary for human survival, the systematic exploitation of plants and animals generated excess wealth that provided a long-term foundation for elite power. By at least 6000 years ago humans themselves were being exploited with the onset of irrigation and metallurgy, which require the grueling labor of ditch-digging and mining.[11] When food is no longer freely available to migratory tribes, he who controls the grainery controls the people.

As historian Lewis Mumford puts it, "the goods of civilization have been achieved and preserved — and this is its crowning contradiction — largely by methodical compulsion and regimentation, backed by a flourish of force. In that sense, civilization is one long affront to human dignity." No longer a shared communal function, "labor becomes degraded into a purchasable commodity." Alongside the unprecedented autonomy of the ruler comes the submissiveness of the farmer and the brick-layer.[12]

The unnoticed side effect of domesticating grains and herd animals was the domestication of ourselves. If the free ranging human tribe is *wildtype*, the new-and-improved personality is *mutant*. What Milgram discovered is not that our nature is to blindly obey but that wildtypes, once 100% of the human race, now comprise a minority of holdouts. The more we conform — the more we allow corporations and cops and experts tell us what to consume and how to behave and think — the further our wildtype heritage recedes.

In the mutant (or "latent") mindset, we see only existing power structures, not the possibilities that lie beyond. The mutant mainstream press is only following its inverted imperative when it shields the powerful from criticism and the public from the facts while marginalizing movements for justice and freedom. Where wildtype seeks to humanize society, in the mutant position we settle for cognitive therapy and pharmaceuticals designed to make us content in the midst of an increasingly inhuman world.

As Chomsky has pointed out, for years after the slaughter of Vietnam, more than 60% of respondents agreed that the attack was a mistake not just strategically but morally. Findings like this, however, must be taken with a grain of salt. When Milgram polled people about what they would do if instructed to deliver electric shocks to others who failed to correctly answer questions, all the respondents said they would refuse, though we know most of them would have obeyed.[13]

What we tell a surveyor on the phone is very different from how we react under the pressure of authority. Recent polls tell us most Americans favor reducing crime in the ghetto by expanding social investments in jobs and education rather than increased policing and imprisonment, but we still turn out en masse, come election time, to choose between a pair of candidates outdoing each other in tough-on-crime rhetoric.[14]

After World War II civilian markets were supposed to bail out military contractors. Instead of bombers and radar units and tanks, they'd make civilian aircraft and dishwashing machines and stainless steel coffins. But too many companies were chasing the same business, so big money began pressuring the Truman administration to ramp up military production. Then one day in 1948, Secretary of State James Forrestal announced that the Soviets were poised to launch a surprise attack against Western Europe. In reality Stalin was so anxious to prevent another war that he instructed communist parties in the West to refrain from revolutionary activity and even refused to assist Greeks in their resistance to a US-backed fascist coup. Yet the American public by and large bought into the scam, agreeing to funnel tax dollars to underemployed war profiteers. As a result Pentagon spending on procurement increased nearly six-fold in a single year, and it's been jacked up ever since, long after the collapse of the Soviet Union.[15] It seems we'd rather pay for weapons that serve no purpose than question our leaders.

Curiously, of the subjects who went to the end of the board in Milgram's experiment, not all of them felt remorse. These were people who never surpassed the emotional development of a child, unable to recognize their destructive behavior as such but either rationalizing it away or dismissing it altogether. Unlike Braverman, who regretted his action and came out of the experience with a renewed respect for life, Mrs. Rosenblum couldn't reconcile her behavior with a self-image cemented into place since early childhood. Instead of processing regret over her action, she just kept talking about how caring and soft she was. Though the unconscious is the great teacher, regressives cut themselves off from their source of wisdom in a bubble of self-serving abstraction.

According to social philosopher C Fred Alford, Milgram overlooked a crucial factor. Alford points to several cases of uncontrollable giggling, not just Braverman's, and he finds in each outburst explication of the secret desire to impose harm, a desire that stems from buried dread.[16] To impose suffering onto another is to gain temporary relief from anxiety. This factor is indeed important in understanding evil, but it's not the dread itself but what you do with it that counts. When dread is excavated and processed, the result is freedom and compassion. Only when systematically repressed does it generate malevolence.

In contrast to regressives, whose primitive ego locks away primal fear and short circuits empathy, those on the mutant-wildtype spectrum frequently relinquish their own points of view for those of others. Trouble is, in the mutant mindset we prefer empathizing with "winners" over "losers." Thus Alford finds that when most people are confronted with the story of Adolf Eichmann, chief bureaucrat of the holocaust, they would rather rationalize his actions — seeing the world from his point of view — than empathize with his many victims.[17] He was just following orders, and don't we all blindly follow orders?

Mutants are simply wildtypes who, under certain conditions, snap into position, taking on socially-designated dehumanized roles morphed to perfection over thousands of years. No one is devoid of mutant sensibilities, and everyone retains at least a trace of wildtype. Within a lifetime poles can shift, usually from wildtype to mutant though occasionally vice versa. But at least those who lean towards the mutant are emotionally evolved to the point where they can process their own pain. Only regressives chronically evacuate their pain onto others. The mutant majority is a problem only to the extent of its vulnerability to manipulation by the pathologically self-involved.

Where regressives are crazy as individuals, mutants merely participate in autodelusion. Mass mutancy leverages the specific insanity of the regressive into the general insanity of society. Without an army of mutants backing them up, paranoid regressives like Hitler or Nixon could never have attained power, no one would have followed Jim Jones to Guyana and George Zimmerman would never have amassed a tidy legal fund. In a world without mutants, "recording artists" couldn't keep making millions long after they've stopped making music.

Wildtypes and regressives share a predilection for breaking the rules, but the motivation is different. Where wildtypes defy rules that stand in opposition to reason, regressives defy whatever limits their gratification, their control, their status, their wealth, etc. If Nietzsche's superman is the classic model of wildtype personality, the regressive ideal is Ayn Rand's sociopathic misinterpretation thereof.

The mutant-wildtype-regressive dynamic plays out nicely in a *Star Trek* episode where Captain Kirk, the nerd's wildtype ideal, is falsely imprisoned on the planet where the galaxy's remaining lunatics have been deposited. One of them, a chameleon, takes on the appearance of Kirk so as to fool Spock, that pointy-eared rule-bound mutant, into killing the real Kirk. Spock figures out who's the fraud when Kirk tells him to shoot them both to prevent the chameleon from taking over the Enterprise.[18]

The fate of the world depends on whether mutants can step out from under the spell of regressives.

The one political philosophy that makes sense of Milgram's findings is *anarchism*, which seeks to restore perennial human social organization in the context of contemporary technically-advanced life. In opposition to power systems of every kind, whether criminal, corporate or governmental, anarchism is self-governance at every level of structure from family to town to world. Blocking monopolization of power not only keeps regressives from taking command but cultivates the wildtype in all of us. Like tribal society, no division exists between public and private, as we the people care for ourselves and each other without the need for a nanny state or a rich investor to come along and rescue us. In power-centric social arrangements, by contrast, wildtype is punished and repressed, either aggressively as in totalitarian societies or passively as in "free" societies.

The seeming impossibility of anarchism is belied by the fact that the universe itself emerged without authority or design. No center of power, whether theological or mathematical, decreed that photons would separate from atoms, that planets would coalesce from remnants of solar placenta and that oceans would broadcast invisible streams of vapor into the sky to rain down upon pools of primordial chemical feedback loops. Life is a local self-organized system playing out against the backdrop of cosmic self-organization. Whether we call it anarchism or democracy or Taoism, the most sensible mode of human organization is nature's way.

Because living systems resonate with similar predecessors, the more we blindly obey, the more powerful is the general will to subordination. Though wildtype far longer than mutant, we've multiplied considerably since the prehistoric inversion, adding weight to the collective directive.

Despite occasional wildtype resurgence — from the rise of philosophy and democracy in ancient Athens to the Benedictine monasteries that formed the backbone of medieval society to the Renaissance, the Enlightenment and the cultural combustion driving the 1890s and 1960s — we have yet to find our way back to ourselves. In a world where regressives rule and mutants obey, where anarchism is mistaken for mere anarchy and anarchists conflated with the Antichrist, where capitalism — the systematic siphoning of wealth from those who work to those who don't — is called *socialism* by vast hordes of ignorant consumer-barbarians, how can we hope to escape the death grip of concentrated power?

The experiment requires that you continue.

1. Milgram, Stanley, *Obedience to Authority*, New York: Harper & Row, 1974, pp 4, 21, 52-53
2. *Ibid*, p 35
3. *Ibid*, p 72
4. Chazan, David, "Row over 'torture' on French TV," BBC News, March 18, 2010. Available from: http://news.bbc.co.uk/2/hi/europe/8573755.stm (Accessed Jan 3 2014)
5. Milgram, p 79
6. *Ibid*, p 188
7. *Ibid*, p 77
8. *Ibid*, p 9
9. Patel, Raj, *Stuffed and Starved*, New York: Melville House, 2007, p 293
10. Milgram, p 124
11. Toynbee, Arnold, *A Study of History: A New Edition Revised and Abridged by the Author and Jane Caplan*, New York: Weathervane Books, 1972, p 51
12. Mumford, Lewis, *The Transformations of Man*, New York: Harper & Row, 1956, pp 38-39
13. Milgram, p 28
14. Henwood, Doug, "White people can surprise you sometimes," *LBO News*, October 31, 2011 Available from: http://lbo-news.com/2011/10/31/white-people-can-surprise-you-sometimes (Accessed Jan 3 2014)
15. Cockburn, Andrew, "'Follow the Money' Why the U.S. Defense Budget Soars, Even as Military Shrinks," *CounterPunch*, Vol. 18, No. 6, May 16-31, 2011, p 2
16. Alford, C Fred, *What Evil Means to Us*, Ithaca: Cornell University Press, 1997, p 26
17. *Ibid*, pp 74-80

18. Erwin, Lee, "Whom Gods Destroy," *Star Trek*, originally broadcast January 3, 1969. Available from: http://en.memory-alpha.org/wiki/Whom_Gods_Destroy_%28episode%29 (Accessed Jan 3 2014)

Chapter 11

Sympathy for the Dubya

Just out from the birth canal, your retina bathing under their first blast of light, you find yourself in the embrace of something alien. As doom percolates up from your guttyworks to tickle the base of your throat, you squirm desperately to free yourself. Already the truth is wordlessly dawning.

Mommy is Barbara Bush.

Starting in 2001 and continuing for the next eight surreal years, the president of the United States had all the introspection of a weasel, though without the admirable adherence to the natural way of its kind. George W Bush used the power at his command to impose incalculable suffering onto a world he had no capacity to understand. In his pacified intellect, his arrogance and paranoia, his systematically warped view of self and other, he served as both living model and enforcer-in-chief of a nation's psychopathology.

Our Virgil in this unsettling dreamscape is psychiatrist Justin Frank. Author of the justly renowned *Bush on the Couch*, Frank charts W's trajectory from infant to infantile, from backyard bully to global terrorist.

According to Freudian theorist Melanie Klein, each of us is confronted early in life with a schism. The infant regards himself as innately good, while others achieve grace only insofar as they cater to his needs. Only with nurturing does the black and white mindset break down as the child discovers he's still worthy of love despite shameful destructive impulses. With motherly love the child learns he's a mixture of good and bad, same as everyone else, and over time "internalizes the maternal function," making peace with the dark side on his own.[1]

But when your mother is a repressed intimidating control

freak who fails to respond to your emotional needs, the schism between good and bad cements into place. Stuck for life at a primitive stage of emotional development, your world "peopled with unreal figures, uncomplicated by ambiguity," friends and family serve as extensions of the self, while everyone else is the other.[2] After 9/11 Bush articulated the attitude of the schizoid position. "There are no shades of gray in this fight for civilization … Either you're with the United States of America, or you're against the United States of America."[3]

Bush never learned to integrate his fear and rage and guilt into his sense of self, never healed the schism between who he is inside and the affable Joe Sixpack self-image he projects for public viewing. Instead of accepting and processing negative emotions and thereby leaching them of power, he walls them off. Digging up the deadweight of past transgressions would only allow the dam to crack. The longer that barrier stays up, the more mammoth the sea of negativity behind it and the more terrifying the prospect of drowning should it be allowed to open. Bush "denies his fallibility, vulnerability, and responsibility," writes Frank, "because on a fundamental unconscious level he feels he must do so to survive."[4]

"I don't spend a lot of time looking in the mirror," Bush once said, "except when I comb my hair."[5] Never one to reflect, he displaces his malignancy onto his enemies, locating in them the traits that disfigure his own psyche. The enemy-other isn't a human being so much as a cartoon character, a screen onto which his own aggression is projected. There's no shame in brutalizing a caricature.

The normal orientation of the mind, which Klein calls the depressive position, is perceived by regressives like Bush as weakness, though of course it's just humanness, the ability to honestly perceive one's vulnerabilities and character flaws and to process the pain of loss and failure. To empathize with someone else's pain, you must first feel your own. By allowing the integration of negative emotions, sorrow can open the heart and heal a fractured worldview.

When his sister Robin died, little W didn't know any better than to suppress his pain. Children learn to mourn by watching their parents, and George Sr. and Barbara went golfing the day after their daughter's funeral. Fast-forward to 9/11 and big W has no idea how to respond, skipping directly from stunned glassy-eyed incomprehension to fantasies of revenge, meanwhile exhorting his fellow Americans to act like nothing happened and go shopping.

Frank gives numerous examples of Bush's wall of denial, including his refusal to take responsibility for the failure to prevent 9/11, even falsely claiming he'd received no warnings of an impending attack, and his insistence that the Iraqi "weapons of mass destruction" debacle was due to bad intelligence, though of course the data were distorted so as to meet the administration's own specifications. When he says, "Free nations don't develop weapons of mass destruction," you get the feeling he believes it himself.[6] Deep in delusion he tells lies easily exposed, like his claim to Kofi Annan that the US offered one last chance for Saddam to accept UN inspectors before the invasion and his public claim that he couldn't return to Washington after 9/11 because Air Force One had been targeted.

He lied about his gubernatorial record, his campaign opponents and his rationale for war, and he gave Orwellian names to his policies like "Clear Skies," "Healthy Forests" and the "Patriot Act." Despite denials about DUI arrests, illegal money-making ventures, desertion of his National Guard unit during the war against Vietnam, etc., he said with apparent sincerity, "I have been very candid about my past."[7] He set new records in the nation's budget deficit yet campaigned for a second term as a "fiscal conservative."[8] Though he defied the US Constitution, the Geneva Conventions and the UN Charter, still he could say with a straight face, "We have a responsibility to respect the law."[9]

No matter how many servings of death and misery he ladled out to the world, Bush's sense of his innate wisdom and goodness could not be shaken. Like the psych ward patient who thought he was dead and announced, "Dead men do bleed" when a doctor pricked his skin, the boy emperor thought he was a loving servant of Jesus, a "compassionate conservative" standing against "those who deliberately take the lives of men and women and children without mercy or shame."[10]

Maintaining industrial strength delusion comes at a price: the constant threat that reality will flood back in. Since talking can serve to manage anxiety, language problems often indicate unresolved fears. As Mark Crispin Miller points out, W's "dyslexicon" doesn't manifest when he speaks punitively but only when trumpeting the qualities that reinforce his delusional self-image, such as idealism, altruism, compassion, etc. His alcoholism, according to Frank, stemmed from a failed effort at anxiety management. Once his drinking got out of control and became a source of anxiety in itself, he turned to religion, giving his life over to his god in exchange for a sense of certainty and omnipotence. His rigid exercise routine serves the dual purpose of maintaining a comforting sense of order and bulking up his body to hide inner weakness.

While the phallic stage of development is ordinarily overcome by age seven, his frequently absent father couldn't help him develop autonomy from his overbearing mother and her self-described "beautiful mind." In the White House he could fool himself and a few others with his erect posture, bulging biceps and "bring 'em on" bravado, but the rest of the world saw a scared little boy hopelessly dependent on his substitute president, Dick Cheney, and his substitute mom — that other *bush on the couch* — Laura. Whereas most phallic narcissists hope only to one-up the neighbors, W "has the entire US military to function as his penis, and he swings it around with fierce power."[11]

Humiliated by his unfulfilled need for love and approval from his father, who set unattainably high standards as athlete, fighter pilot and businessman, Bush lashes out at the vulnerable, says Frank, because they remind him of his own unacceptable weakness. Thus we see him detonating fireworks in frogs as a child and, as governor of Texas, gleefully mocking death row inmate Karla Faye Tucker for daring to request a pardon. "Please don't kill me."[12] Frank attributes the infamous smirk to the pleasure Bush takes at imposing onto others the pain he can't process himself. Occasionally it even spills over into giggles, like when he was asked how he could send to death someone whose lawyer slept through the trial.

Bush's specialty in office was raising hopes before sticking the knife in. Only after praising Americorp and promising 50% more money did he cut its budget by 80%. Same thing with school kids and even New York firemen. He made sure to tease poorly outfitted troops with promises of excellent healthcare before cutting benefits and making them wait months, in many cases, to get treatment for wounds suffered in the course of his war of conquest.[13]

Frank sees Bush's refusal to allow photographing of flag-draped coffins as a sign of his contempt for soldiers who demonstrated precisely the kind of bravery he himself lacked during the Vietnam era. His central imperative is to protect his self-image of virility and moral good sense against anyone who might cast them into doubt, including nosy reporters with their "trick questions." When delusions of grandeur stem from a hidden core of paranoid insecurity, the diagnosis is megalomania.[14]

Bush's fear-driven need for power gave us the "unitary executive," a virtual dictator compulsively appending "signing statements" to laws passed by Congress so as to exempt him from recognizing them. When Congress banned abuse of military prisoners, he issued a statement to the effect that he could ignore the legislation whenever he pleased.[15] Meanwhile he granted himself the power to declare anyone in the world, including Americans, "enemy combatants" who could be imprisoned and tortured and even executed without charge. When this astounding attack on the Constitution went before an appellate court, the panel of judges agreed with Bush on the grounds that an enemy combatant is a "nonperson" and therefore has no rights. In gratitude for this betrayal of fundamental human values, he nominated one of the judges, John Roberts, to the Supreme Court within days of this ruling.[16]

In the schizoid position, anyone who doesn't consistently praise you must be attacking you. As former Treasury Secretary Paul O'Neill found out, Bush needed to be told he was right regardless of the reality. "Loyalty to a person and whatever they say or do, that's the opposite of real loyalty, which is loyalty based on inquiry and telling someone what you really think and feel — your best estimation of the truth instead of what they want to hear."[17] Unwilling to bow down to the Texas tyrant, O'Neill quit.

Drunk on ego, W went to Egypt and told Arab leaders he would bring peace to Palestine. "Look guys," he said, "if I didn't think I could do this, I wouldn't be here. I wouldn't waste my time and come all this distance if I didn't know I could do it."[18] His repeated reference to an imagined "axis of evil" implied that without him in the Oval Office we'd all be facing Armageddon. His greatest ally was none other than Osama bin Laden, as the more destructive the villain, the more glorious the protector. Only in the unfeeling unreality of the schizoid position could Bush exploit a tragedy like 9/11 to manipulate his country into imperial war.

Bush's followers, whether regressive or merely mutant, bought into his savior routine as much as he did. His "shock and awe" campaign in Iraq catered to the infantile need for fantasies of omnipotence in the face of helplessness. His assertion that terrorists "hate our freedoms" resonated with people who prefer self-serving delusion in place of real-world complexity. Like children newly aware of their destructive impulses, his constituents projected their lower nature onto an enemy while identifying with an idealized father-hero.[19] Only cult-like admirers could fail to see the irony in draft-dodging W calling suicide bombers cowards.

Freud would have classified Bush as an "Exception," someone who thinks he's above the law, entitled to get whatever he wants without explanation or apology.[20] Frank's book, published prior to the 2004 election, served as a plea for Bush's removal from the presidency, an office which could only exacerbate his megalomania. But he held on, in part, because he reflected back onto us our collective narcissism, commonly known as *American exceptionalism*. As Bush thinks he's America, America thinks it's the world.

W often posed the rhetorical question of why the terrorists hate us. Frank zeroes in on the real mystery: why does he hate us? Surely a person privileged from birth – handed the White House as casually as the car keys on a Friday night – is happy with his lot in life. What could ordinary Americans possess that makes W despise us such that he routinely lied to us, teased us with unkept promises, brought us to the brink of financial meltdown and mired us in horrifying wars of aggression?

Could it be that he hates us for our freedom from schizoia? Far easier to integrate your negative aspects into your self-understanding and attain a measure of peace than project your evil onto the enemy-other and expend energy holding back the anxiety that the truth will one day emerge. This anxiety can only be enhanced when the cartoon evil of your own creation comes to life and hurls exploding airplanes at your chief symbols of wealth and power. W had demons attacking him from within and without. Like a corporation imposing health and environmental costs onto the public, he externalized his private hell onto Afghanistan and Iraq.

Yet it was President Bush, not Obama, who pulled the plug on the war against Iraq. That he ended it in abject failure, with no military presence and no preferential treatment for US oil companies, indicates that the whole escapade was really about dressing up in a fighter pilot costume and declaring "mission accomplished" on an aircraft carrier. Since his dad had left the job in Iraq undone — the hatchet not yet driven into the forehead of the nation — the implicit message was that he, not Daddy, was the real hero after all.

By signing the Status of Forces Agreement in December of 2008, Bush locked the US into withdrawing all troops from the country within three years. Obama did his utmost to wiggle out from the arrangement so US troops could continue occupying, at taxpayer's expense, an archipelago of military bases that should never have been built in the first place.[21] Fortunately the Iraqi government rebuffed him, refusing to accept legal immunity for US troops after learning of an unprosecuted 2006 massacre of civilians, including children, from Wikileaks by way of a real hero, Chelsea Manning.

Perhaps to counter the loss of Iraq, Obama bombed another oil rich Arab nation, Libya, on the fabricated pretext of preventing genocide. As with Bill Clinton's lethal assault against the former Yugoslavia, diplomacy was never genuinely pursued despite willingness to negotiate from the other side. Whether Kosovo or Libya, the result of military force was lots of dead civilians and the insertion of Western influence into a formerly autonomous region.

Electing a liberal to the White House has never diminished the US drive to world dominance. What we have is a power system with two wings, Democrat and Republican. Elite liberals portray bombings as opportunities to promote democracy and human rights, while their counterparts on the right rave about the windfall of cheap resources to fuel corporate profits back home. Obama is every bit as subservient to both Wall Street and the national insecurity state as his predecessor. The left-right schism serves to distract the people as power continues concentrating.

Bush was able to start two wars not just because of 9/11 but because previous Democratic administrations had already laid the groundwork in both nations.

Years after the fact, Jimmy Carter's National Security Advisor Zbigniew Brzezinski revealed that US military aid began flowing to Afghanistan in the summer of 1979, six months before war broke out, "to induce a Soviet military intervention." Even US officials have described the fruits of his efforts as "indescribable horror." But when he was asked if he felt any regrets, Brzezinski seemed puzzled. "Regret what? The secret operation was an excellent idea. It had the effect of drawing the Russians into the Afghan trap, and you want me to regret it?"[22]

When confronted with the fact that at least half a million Iraqi children died as a result of the bombing of Iraq's infrastructure and the subsequent years of sanctions, Madeleine Albright infamously retorted, "we think the price is worth it," and was soon thereafter promoted to Secretary of State.[23] At any time during his presidency, Bill Clinton could have normalized relations with Iraq, rendering W's subsequent invasion politically impossible. He was even informed by a high-level defector in 1995 that Iraq had long-since destroyed its WMDs, and still he failed to change course. Even after the invasion, the Democratic-controlled Congress could have cut off funding for the occupation. As with the arming of the mujahedin, the devastation of Iraq was bipartisan.

And let's not forget Bush's buddies at the *New York Times* and the rest of the liberal media establishment who not only promoted the phony WMD threat prior to the invasion but fed the public the big lie that the occupying force was the legitimate authority in Iraq while popular resistance was an "insurgency," as if the country legally belonged to the US. Meanwhile his pals at the "socialist" UN failed to enforce its own charter and condemn his crime of aggression, which would have at least provided Iraqis with some diplomatic breathing room.[24] By invading their country and neutralizing the UN, the US invited Iraqis to take up arms and then called them terrorists for doing so.

Since resistance expressed widespread revulsion at US occupation, there was a certain logic to attacking whole cities, issuing orders to "kill all military-age males" during certain operations and calling in airstrikes over heavily populated urban areas.[25] In the end 10% of the Sunni population of Iraq was wiped out, not just in the course of sieges conducted by foreign troops but at the hands of Special Police Commandos organized by US officials who then blamed paramilitary violence on sectarianism.[26]

Just three months after invasion, the occupying authority pulled the plug on a citizen-led effort to organize local elections, leaving Saddam's Baath Party officials in power and giving the lie to the claim – after the dreaded WMD failed to materialize – that the invasion was about bringing democracy to Iraq.[27] Since the only alternatives allowed to compete against Western-backed candidates were Islamists, in a country long accustomed to secular government the imams predictably lost.[28] When the Arab Spring blossomed in Iraq, the US-sponsored Iraqi government reacted exactly like other repressive regimes of the region: shooting, arresting and torturing protestors.[29]

Not merely the mad scheme of a lone megalomaniac, the invasions of Afghanistan and Iraq were expressions of a schizoid culture that projects its true nature onto its enemies while regarding itself as the "city on the hill," bearer of freedom and prosperity to all the world's people.

Despite its origins in rebellion against empire, the imperialist meme found fertile soil in North America. The quest for overseas power was in full swing by the 1890s, starting with Hawaii in '93 and carrying on to the Caribbean and the Philippines five years later. In the 1920s alone, US troops intervened on behalf of big money and in opposition to collective advancement in Nicaragua, Haiti, the Dominican Republic, Cuba, Russia, Panama, Guatemala, Turkey, China and Honduras. Many of these operations included bombings and long-term occupations.[30]

But all this was child's play compared to the murderous assaults during and after World War II. While Americans typically believe the nuclear obliterations of Hiroshima and Nagasaki in August of 1945 forestalled a ground invasion of Japan, the war was effectively over in the spring of that year when Emperor Hirohito asked the Soviets to mediate negotiations for surrender to US forces. President Truman was aware of Japanese desperation to end the war but refused to budge from his demand for unconditional surrender, meaning that Hirohito would have to abdicate the throne.[31] Yet once the US occupation was in place, the emperor was allowed to stay on after all. Thus the most spectacular acts of mass murder in world history served no purpose. No wonder historian William Blum calls US foreign policy "clinically mad."[32]

After defeating Japan, the US felt entitled to its "protectorate," Korea. Only the Soviet Union stood in the way. So the US divided the country with the Soviets along the 38th parallel on the understanding that elections would be held to reunify the country. Instead the US suppressed the grassroots provisional government in the South and held a sham election that empowered American-educated strongman, Syngman Rhee, who stacked his regime with collaborators of the Japanese occupation. According to a US diplomat serving in Korea, Rhee executed "probably over 100,000" dissidents without trial. The war, which was ongoing from the moment Rhee took power, was an assault against the Korean people by a deranged US-backed regime.[33] Bombing of northern Korea was more intensive than anything in WWII and finally stopped only when "intelligence" could no longer find any targets higher than one story. By the time it was over in 1953, four million people had been killed.[34]

As Blum points out, to this day the war is said to have started on June 25th 1950 when the North invaded the South. Yet this action followed two days of bombing from the South. Rhee regularly announced his intention to take the North by force, and the two sides had been skirmishing for years.[35] The only factor that distinguished the North's action of June 25th was its effectiveness, solidifying the power of megalomaniac Kim Il-Sung, whose disastrous ascent was a predictable byproduct of US provocation.

That the war against Korea was blamed on the northern part of Korea itself just goes to show the power of anticommunist autodelusion. In the popular mind, the entire struggle was reduced to the claim that the country had to be divided so the Russians didn't take the whole thing. Yet the Russians, from their point of view, went along to prevent the US from taking it all. That the "other" even has a point of view never enters the narcissistic mindset.

Without messing a beat, the US took command of Vietnam from the French in 1954 with the understanding, once again, that power would be handed over to a duly elected government.[36] America's promise to the Vietnamese people was broken once it became clear that they were determined to elect a communist, Ho Chi Minh. The subsequent war against democracy in Southeast Asia cost at least two million lives.[37] Since no major economic interest was at stake, as in Korea the basis for assault was paranoia that communists were bent on conquering and enslaving the world, pretty much what the West had been doing for centuries.

After *US News & World Report* called the war "the most significant example of philanthropy extended by one people to another that we have seen in our times," Blum sent a pamphlet to editor David Lawrence detailing atrocities committed by US troops in Vietnam. Lawrence responded that the pamphlet only proved his point, "namely that primitive peoples with savagery in their hearts have to be helped to understand the true basis of a civilized existence."[38]

It was during this era of full spectrum derangement that W came of age. As president he effortlessly matched his inner drives with the public task of promoting power and punishing the powerless, all the while imagining he stood for everything good and decent and civilized. The schizoid culture that generated him was in turn reinforced by his rule.

Everyone regresses under stress. We've all sometimes projected our shameful impulses instead of confronting them. It's pathological only when it becomes ingrained, when it's no longer a personal choice but a function of biology, of nature's memory.

Just as America was drawn into the groove established by empires dating back thousands of years, in his youth George W Bush followed helplessly the developmental trail blazed by megalomaniacs of old. All those emperors and popes and fascists who preceded him were implicit in him as he explicated them. Due to initial similarities in genetics and upbringing, he came into resonance with his malignant forebears, a resonance that increased with each step on the road to schizoia.

1. Frank, Justin, *Bush on the Couch*, New York: Regan Books, 2004, p 10
2. *Ibid*, p 11
3. *Ibid*, p 13
4. *Ibid*, p 13
5. Allen, Mike, "Frustration Nation," *Time*, September 4, 2006, p 35
6. Frank, p 200
7. *Ibid*, p 90

8. Newton-Small, Jay, "Daring Bush to Veto," *Time*, October 29, 2007, p 13
9. Frank, p 92
10. *Ibid*, p 136
11. *Ibid*, p 146
12. *Ibid*, pp 101, 111
13. *Ibid*, p 200
14. *Ibid*, pp 108, 133
15. Sullivan, Andrew, "We Don't Need a New King George," *Time*, January 23, 2006, p 74
16. Floyd, Chris, "Roberts Voted to Give Bush Dictatorial Powers," *Anderson Valley Advertiser*, August 24, 2005
17. Dickerson, John, "Confessions of a White House Insider," *Time*, January 19, 2004, p 36
18. Carney, James, "How Bush Got Religion," *Time*, June 16, 2003, p 42
19. Frank, pp 129, 173
20. *Ibid*, p 78
21. Hirthler, Jason, "Delusions of the Liberal Intelligentsia: 10 Myths About Obama and the Democrats," *counterpunch.org*, November 13, 2012. Available from: http://www.counterpunch.org/2012/11/13/10-myths-about-obama-and-the-democrats/ (Accessed Jan 3 2014)
22. Blum, William, *Rogue State: A Guide to the World's Only Superpower*, Monroe, Maine: Common Courage Press, 2000, pp 4-5
23. *Ibid*, p 6
24. Davies, Nicolas J.S., "The Victory of Popular Resistance in Occupied Iraq" (Part One) *Z Magazine*, December 2011, p 31
25. *Ibid*, p 33
26. Davies, Nicolas J.S., "The Victory of Popular Resistance in Occupied Iraq," (Part Two) *Z Magazine*, January 2012, p 25
27. Davies, Nicolas J.S., 2011, p 32
28. Davies, Nicolas J.S., 2012, p 23

29. *Ibid*, p 28
30. Sardar, Ziauddin and Davies, Merryl Wyn, *Why Do People Hate America?* New York: Disinformation, 2002, p 95
31. Wainstock, Dennis, *The Decision to Drop the Atomic Bomb: Hiroshima and Nagasaki August 1945*, New York: Enigma Books, 2011, pp 168-169
32. Blum, p 26
33. Blum, William, *Killing Hope: US Military and CIA Interventions Since World War II*, Monroe, Maine: Common Courage Press, 1995, pp 45, 49-54
34. Smith, Stansfield, "North Korea's Justifiable Anger," *counterpunch.org*, April 10, 2013. Available from: http://www.counterpunch.org/2013/04/10/north-koreas-justifiable-anger/
35. Blum, 1995, pp 46-47
36. "The Final Declaration of the Geneva Conference: On Restoring Peace in Indochina, July 21, 1954," *Modern History Sourcebook*. Available from: http://www.fordham.edu/halsall/mod/1954-geneva-indochina.html (Accessed Jan 3 2014)
37. Sardar and Davies, pp 97-98
38. Blum, 1995, pp 8-9

Chapter 12

The Moloch Whisperer

On a sunny morning in September of 2001, as the glass and steel façade of the World Trade Center peeled away, the American people caught a glimpse of reality. From the initial shocked silence, there came an opening for discussion and collective self-critique. Voices were raised about the culpability of a country so fixated on its own security that it routinely defiles that of its neighbors. But the opportunity was snuffed out in the rush to war. To call up our history as a rogue terrorist state seemed in bad taste. Better not "attack" the homeland when it's at war.

Like the Titanic, whose pilot thought he'd look like a sissy if he didn't plow through that ice field, the Twin Towers went down under the weight of bloated ego. Plunged into oceanic abyss, America dutifully crawled into the fetal position, preferring to float in the debris of its own hallucinatory waste than come to grips with the true cause of 9/11: payback for decades of lethal intervention in the world's largest oil-extracting region.

That the public would acquiescence so easily to vicious reprisals baffled many on the left. Observers typically blamed the narcotizing power of mass media, failing to recognize how the media's messages are shaped, in part, by the masses themselves. Once again the role of PR was overrated, the will of the people to remain comfortably ignorant overlooked. That's not to say the problem is "the people" rather than "the elite." Quite the contrary, the problem is collective insanity.

Following the 2003 capture of Baghdad, none of the major networks reported the widespread use of cluster bombs and depleted uranium shells, which slice up armored vehicles and expose anyone in the vicinity, including US soldiers, to radiation. While the rest of the world got a reasonably clear picture of the human costs of the invasion, Americans were kept in the dark by all-powerful media. For this reason, argues Norman Solomon, W's popularity soared at home exactly as it was crashing in the rest of the world.[1]

Or maybe Americans by and large figured Saddam got what he had coming. Maybe those "funny foreigners" didn't get it because they're not on Team USA. Solomon himself notes the Pentagon's deck of cards featuring Iraqi bad guys to be hunted down by American good guys was hugely popular. Given the opportunity to feel vicariously heroic, most people duly rewarded W with approval points.

Sheldon Rampton and John Stauber, authors of *Weapons of Mass Deception*, point out that every network in the US refused to air antiwar ads during the invasion. Does this mean nationalist ideology is bigger than money on the corporate airwaves, or did the networks figure the ads weren't worth the money they brought in if they turned off viewers? Rampton and Stauber also cite MSNBC's dismissal of Phil Donahue prior to the invasion as an example of pro-war bias, noting that Donahue's show attracted more viewers than any other program on the network.[2] But if the frequency of his anti-war guests triggered the taboo of "anti-Americanism," the result might have been a catastrophic loss of viewership. Donahue himself cites the "tremendous fear corporate media had for being on an unpopular side during the ramp-up for a war," invoking General Electric's ties to the defense industry only as an afterthought.[3] What motivates the media is not so much the desire to manipulate people as the abject fear of alienating them.

In *Time* magazine Michael Elliot characterized the US objective in Iraq as the transformation of a "wolf into a lamb." Yet Iraq long ago abandoned Kuwait; this time the US was the one sinking its teeth into prey. In passing, Elliot claimed the only legal occupiers of Iraq were the US and Britain, ignoring the fact that invading another country is the ultimate war crime, a violation of both the Geneva Conventions and the UN Charter. The occupation of Iraq was about as justifiable as Hitler's seizure of Poland. Were Elliot's distortions the result of elite conspiracy to confuse the American public, or was he just another confused American himself? In a sidebar to Elliot's story, Al-Qaeda operatives were called "extremists" and "God-obsessed terrorists," which is true, though it's hard to imagine a more apt description of the Bush administration.[4] Was *Time* playing the public, or were its staff blinded by the same unreflective narcissism infecting the country as a whole?

This is not to deny that corporate media outlets routinely skew issues in favor of elite interests. More importantly though is how ordinary people actively identify with empire. Americans typically have come to expect continued supremacy, and anything that seems to threaten that power sends ripples of anxiety across the public mind.

According to a poll conducted by CBS and *The New York Times* prior to the invasion of Iraq, a clear majority of Americans thought Saddam was personally involved in the 9/11 attacks, that Iraqis were aboard those planes and that terrorists back in Baghdad were planning more such attacks.[5] These beliefs seem to have emerged out of thin air. Not just passive receptors of propaganda, Americans wove it themselves from the most threadbare of fabrications.

During the Gulf Massacre of 1991, US bombing of Iraq's sewage and water treatment plants unleashed an army of homegrown germs against its citizens. Upwards of a million Iraqis, mostly small children, perished in the biological onslaught. The death toll was common knowledge. Yet Americans, by and large, felt no qualms about invading an already bombed-out, blockaded country over the hypothetical possibility that it was stockpiling deadly chemicals and germs. So virtuous are we that our actual evil is outweighed by their potential evil.

Despite justifying invasion on the grounds of WMDs, the US felt no compunction to demonstrate that Iraq actually possessed any. On the contrary, Saddam had to prove he didn't have them. As White House spokesperson Ari Fleischer remarked, "I think the burden is on those people who think he didn't have weapons of mass destruction to tell the world where they are." Yet the bizarre illogic of it all went right past the mutant American mindset.[6]

Let's not forget that when Saddam did possess WMD, they were supplied by the US and other Western powers and wielded, with tacit US approval, against Iranian troops and Kurdish Iraqis. The attack on Iran was well underway in 1983 when Saddam received a visit from Donald Rumsfeld bearing a pair of golden spurs for the mad dictator. Purely on the basis of his long-ago possession of toxins and germs *that we gave him*, the US now felt justified in condemning him to death. How could the general public be undisturbed by the idiocy of all this?[7]

Not just W but millions of Americans thought US troops would be welcomed as liberators, ignoring the fact that the bombs falling like raindrops in '91 were all made in the USA. After seeing much of their modern infrastructure bombed to infinity and years of sanctions that prevented rebuilding, after all the suffering they endured, Iraqis were pummeled with more bombs and a full-scale invasion, at which point they were supposed to line up in the streets cheering the troops and blowing them kisses as they rode by. This was fully expected and eagerly anticipated by people all over the country.

What is going on here?

Imagine if the president, every January, had to give a mental state of the union address. An accurate assessment would indicate grandiosity, arrogance, need for admiration coupled with defensiveness, a malignant lack of empathy, the willingness to take advantage of other nations — all of which must automatically comply with our wishes — the belief that our exceptional character exempts us from normal rules of conduct and, finally, the assumption that when others dislike us, it's because they envy us.

These are in fact the diagnostic criteria of narcissistic personality disorder, of which megalomania is the extreme form. The United States is indeed a great nation, but the shining steel exterior conceals a toxic burning need to control and punish, to be obeyed and serviced and flattered. As veteran muckraker Bruce Anderson writes, "some of us need reminding that for all the hype about us being the most advanced country in the world, we're also the most primitive country in the world."[8]

Since our ever-shifting enemies cannot help but see that our way is best, they must oppose us out of hatred. Hence the infantile lie, repeated ad nauseam, that terrorists hate us "because we advance peace and democracy," in the words of Bill Clinton.[9] If Iran is trying to build the Bomb (for which no evidence has ever been produced) its purpose can only be to destroy Israel, though such an act would obviously be suicidal. The real reason, which cannot be voiced, is that nuclear weapons might deter unprovoked US or Israeli aggression on Iranian soil.

A beacon of light, the world's selfless promoter of freedom and prosperity, we expect others to accept a narrative of history centered entirely on our POV. With our psychopathic predilection for lethal violence, we're a danger to ourselves as well as others. In short, America is cracked wide open for all to see.

Insanity is generally regarded as a private affliction, something that strikes the rare individual who must be segregated from an otherwise healthy society, but as often as not it's society that needs help. When the deranged is the norm, the lunatic isn't the lone voice of rage and despair but the mass of people who can't or won't listen.

The vast majority of Americans in no way conform to the profile of a personality disorder, whether narcissistic or paranoid or any other face of schizoia. Most of us go about our daily business on the assumption that we have to earn our living and get along with our neighbors. Though our character as individuals is diametrically opposed to our character as a society, we don't always act according to personal intelligence and conscience. One nation caught in a groove, we can't see beyond the collective fantasy.

At the heart of every narcissistic pathology, whether individual or collective, is insecurity. "It's a frightened country," as Chomsky observed after 9/11. Fear suspends our capacity for sound judgment, leaving us vulnerable to hysterical overreaction. How else could people across the country believe the invasion of a defeated and powerless nation was in self-defense? "Levels of fear here of almost everything — crime, aliens, you pick it — are just off the spectrum." So how did we get this way? Chomsky isn't sure, but he urges us to look for answers in our formative years as a nation. "It probably has to do with the conquest of the continent, where you had to exterminate the native population, [also] slavery, where you had to control a population that was regarded as dangerous, because you never knew when they were going to turn on you." Though we can destroy the world several times over, we can't shake the feeling that "somebody is going to come after us."[10]

"There cannot be the slightest doubt," as Konrad Lorenz once observed, "that human militant enthusiasm evolved out of a communal defense response of our prehuman ancestors."[11] All it takes is a threat grave enough to trip our ancient predator-defense reflex, and individuality vanishes as we all band together to scare off the beast. We're not so different from chimpanzees that scream and jump around at the sight of a prowling tiger, throwing sticks and rocks at it and hugging each other for moral support. After 9/11 the tiger was al-Qaeda. With a wave of the wand, it was Saddam.

The American people aren't the first to fall under the spell of general insanity, a phenomenon as old as civilization. Like Chomsky says, we must look to the past for answers but much further back than he suggests.

In the course of researching the origins of war, Barbara Ehrenreich discovered that until relatively recently people were routinely killed and eaten by predator beasts, especially lions and tigers. Somehow the human race collectively forgot its long-time status as cat food.[12]

Homo sapiens is the only species ever to transform from prey to predator. Though our hominid ancestors relied on meat for the protein required for their big brains, they didn't know how to bring down big game and thus had to acquire their meat mostly from scavenging. According to anthropologist Lewis Binford, it wasn't until about 70,000 years ago that humans learned to hunt, and even then it was more like herding than what we think of today as hunting. Only by the Upper Paleolithic, around 30,000 years ago, did humans become highly skilled hunters. Though the tools they forged were later used in warfare, for millennia the men who wielded these weapons used them only to hunt animals, not each other.[13]

Signs of war in prehistory go back only so far. Prior to 12,000 years ago, we find no drawings of military formations on cave walls or mass graves with spear points embedded in skeletons. Though long thought to be an expression of inherently violent human nature, war is a cultural product with a relatively recent onset. Triggered by a combination of overhunting and climate change at the end of the Ice Age, warfare arose in the wake of widespread devastation of not only grass-eating herd species but the predators that fed on them.[14] The warrior band, writes Lewis Mumford, popped up when the hunting band hit the skids. As Ehrenreich explains in her 1997 book, *Blood Rites: Origins and History of the Passions of War*:

With the decline of wild predator and game populations, there

> would have been little to occupy the males who had specialized in hunting and anti-predator defense, and no well-trodden route to the status of "hero." What saved the hunter-defender male from obsolescence or a life of agricultural toil was the fact that he possessed weapons and the skills to use them. Mumford suggests that the hunter-defender preserved his status by turning to a kind of "protection racket": pay him (with food and social standing) or be subject to *his* predations.[15]

The presence of warriors in other settlements provided a menace for local warriors to defend against and vice versa. According to war historian Gwynne Dyer, "Pre-civilized warfare ... was predominantly a rough male sport for under-employed hunters."[16] For eons hominids and humans were menaced by wild animals. When the threat receded, like an auto-immune disease the tools developed to fight it were turned against our own kind. Instead of stalking us across mountain and savanna, the beast was internalized and began preying on us from within.

The beauty of *Blood Rites* is how Ehrenreich weaves the history of war and religion into a seamless narrative. Religion emerged everywhere in terms of sacrificial offerings to hungry deities. "Sacrifice," says Ehrenreich, "is the means by which humans feed the god or gods; the victim is in fact a meal."[17]

Olympian gods longed for the scent of roasting flesh. Aztec deities faithfully spinning the sky each night were fueled in their labors by human blood, as were the sharks to which ancient Hawaiians made offerings of themselves at periods of heightened anxiety. The oldest known altar, discovered in a cave in Vallon-Pont-d'Arch in France, is a rock on which is placed the skull of a bear. The Hebrew Bible is riddled with testimonies to Jehovah's lust for raw meat. In the beginning the beast was God.[18]

That predators terrorized the human race in no way disqualified them from taking on the status of deities, as they were also essential to our ancestors' survival as scavengers. With the worldwide collapse of herd populations at the end of the Ice Age, a partly-eaten kill would have been a godsend. As Ehrenreich argues, the first blood offerings were probably investments intended to keep predators nearby. To this day the Koyukon people of northwestern Alaska maintain that wolves intentionally leave behind part of their kills for humans, who naturally reciprocate with portions of their own. The native tribes of southern California venerated the mountain lion as a provider of meat. The Cheyenne claimed panthers killed deer for them in gratitude for having been suckled as cubs by Cheyenne women.[19]

Just as the pious both love and fear their deity, at one time people everywhere felt exactly the same ambivalence toward wolves, lions, panthers, etc. The bear that provides for you could just as easily turn on you. The lord giveth and the lord taketh away.

The trauma of ancient predation is still with us. Even in urban areas, the chief fear among small children is wild animals, especially snakes, lions, tigers and bears, though giant hyenas, eagles and wolves also preyed on our ancestors.[20] We see ourselves in the humble California ground squirrel, which still exhibits stereotypical anti-rattlesnake behavior despite being free of rattlesnakes for as long as 300,000 years. When his two-year-old son developed a fear of large animals at the zoo, Darwin wondered if we're haunted by ghosts of predators past.[21]

Long regarded as racist slander, human sacrifice is now known to have occurred throughout the world over thousands of years, "from small-scale tribes to mighty urban civilizations." A man shot by arrows lies buried under the entrance of Stonehenge. Two miles away, at "Woodhenge," archaeologists found a three-year old child whose skull had been split open prior to burial. Somehow we still sense that our lord is hungry, even if its growl is no longer audible. In Paleolithic times, when a tiger suddenly appeared from the brush, someone was going to die. As with any trauma, the anxiety returned like clockwork even after the threat had all but vanished. So we simply did as we'd always done before: sacrifice one that others may live.[22]

Over time, as the trauma began to weaken, the sacrifice of our own kind gave way to animal offerings and even circumcision, in which the foreskin substitutes for the man. Only in war does human sacrifice find a durable institution. Not merely make-work for underemployed hunters, war is a religious mission. To this day the battlefield is sanctified, the soldier who dies in defense of his people a glorious sacrifice. To oppose the insanity of systematic human slaughter is seen as a diabolical affront to "God-fearing" people.[23]

Key to the Ehrenreich thesis is the presence of the past. Given the power of materialist ideology, it's no surprise that Ehrenreich, who holds a PhD in biology, assumes the war psychosis is conveyed down the millennia via genes. With the recognition of memory as an inescapable side effect of absolute presence — a property of nature available to all intrinsic forms — genetic artifice can be dispensed with. No need to appeal to an assemblage of nucleic acids to account for our collective post-traumatic stress.

Species-memory individuates with every infant, providing a roadmap for bodily as well as cognitive development. Due to repetition of attack by wild animals, our species-mind contains a deeply ingrained association between predator attack and sacrifice. The more victims of predation, the stronger the resonance of subsequent victims with their forerunners. A Mayan priest could trigger the resonance by dressing as a jaguar before sacrificing a child. Same goes for warriors going "berserk," which means to act like a bear. The sight of blood draws the predator from its lair in the wilderness of collective memory, generating a sense of godlike power for warriors while unleashing terror in their victims.

Despite our natural propensity to share and cooperate, the right cues prompt the species-mind to cough up a command to slaughter the enemy-other. Conscience is all too easily overridden by the imperative to transmute from prey to predator, like Clark Kent to Superman. Suddenly we're like grasshoppers in a drought switching from harmless insects to a roving storm cloud of annihilation, a.k.a. locusts. Whenever the prospect of mass killing is perceived as normal and even desirable, as in 1914 when a frenzy of nationalism swept across Europe, the beast has awakened. We know this creature well. Like a family pet, it even has a name: *Moloch*.

An abstraction of Moloch's properties has built up in the implicate order as surely as the properties of an atom. Mass sense of being under attack puts populations into resonance with previous peoples in similar circumstances. Everyone knows his role, from general to private. It's no accident soldiers dress in uniforms while marching in formation to a beat. Every army materializes Moloch, every soldier a cell of its body.

Christianity can be seen as a failed attempt to short-circuit the sacralization of war. With God himself offered up as the sacrifice, the logic of blood offering is turned on its head. To love our enemy is to recognize that they are us, that we've been killing our brothers and raping our sisters. The real enemy, though invisible and inaudible, is Moloch clamoring for blood. Our forebears defeated the one in the jungle. At its core Christianity is about slaying the one in the mind.

With war came social dominance for the warrior. Since manhood was equated with the power of predation, woman was demoted to prey. Says Ehrenreich, "By assigning the triumphant-predator status to males alone, humans have helped themselves to 'forget' that nightmarish pre-history in which they were all, male and female, prey to larger, stronger animals." Patriarchy "conveniently obliterates our common past as prey, and states that the predator status is innate and 'natural' — at least to men."[24]

All of us are subject to aftershocks from the trauma of predation, but in men the memory of weakness and vulnerability is more likely to be sealed off and forgotten. When the anxiety begins to emerge from its tomb, as it must, there's nothing like a good killing to re-establish a sense of manhood.

What could more aptly summarize human vulnerability and suffering than a defeated Christ on the cross? Yet the terror of war barely missed a drumbeat with the onset of Christendom. Instead of empathizing with the victim, European states fought incessantly and ultimately embarked on a global rampage, crucifying the Americas and Africa, India and Indonesia.

Killing, plundering and enslaving, the conqueror can always hold his oceanic anxiety at bay. Only the defeated come face to face with the universal condition. Those with power are precisely those who don't learn, while those who understand wield no influence over the course of events. Uncaged, the beast is free to graze on human emotions while grotesquely distorting human perception and culture.

Moloch's spell over the West was cast in the eighth century when Europe faced a far more advanced Islamic civilization. According to Ziauddin Sardar and Merryl Wyn Davies, authors of *Why Do People Hate America?*, "Europe lived and developed within a war psychosis, conscious of an enemy with whom it was not evenly matched." Following the Battle of Tours in 732, when Arab penetration of the European heartland was turned back, the original anti-Muslim polemicist, John of Damascus, began to portray Islam as a demonic travesty of religion designed to encourage aggression and lust.[25]

Whether or not Damascus believed his own lie, it soon took on memetic momentum. Projecting Moloch onto the "other" only brought it to life in Europe's own imagination. The Crusades, which served to renormalize war for Europeans infected with Christianity, were launched in 1095 by Pope Urban

II, who guaranteed a place in Paradise to those who took up the cross against "unclean" Muslims. Four centuries later it was Columbus's promise to open up a new front from the east in the war on Islam that got him his ships.[26]

"The old and familiar reflexes of thought shaped in the medieval power rivalry between Christendom and Islam," write Sardar and Davies, "were extended to manufacture the ideas, attitudes and means of dealing with Native Americans." Like his European predecessor, Homo americanus took shape in opposition to people initially far stronger and, in the context of the New World, more knowledgeable and adept. Yet these "savages" would soon have to step aside as the superior race laid claim to its New Canaan. Once its borders were established, the US began preying on peoples overseas. "As peace, order and prosperity followed our expansion over the land of the Indians," said Theodore Roosevelt in 1898, "so they'll follow us in the Philippines." In a sense he was right. Filipinos were slaughtered at a per capita rate exceeding that of Vietnamese 70 years later.[27]

Distraught by the emerging slaughter in 1964, LBJ wanted out of Vietnam. He sought a way to avoid escalating military action but felt pressured by the American public to be a "man" and stand up to the Russian bear.[28] Like Truman before him, the experiment required him to continue.

Whereas Moloch dug itself in at the very core of the American mind, the "old world" took a different path. After a pair of mass slaughters, Europe no longer saw glory and redemption in war, only derangement. In the face of incipient European awakening, the US kept imperialism alive, lending support to dictators of newly "freed" nations and, where necessary – as in Iran, Guatemala, Congo, Brazil, Chile, etc. – financing coups and installing strongmen to service the needs of capital.

Today the West has come full circle to the original enemy-other. Soon after 9/11, Rich Lowry of *National Review Online* reported "lots of sentiment for nuking Mecca." He advised caution. When Bush declared, "Islam is not the enemy," he was soundly thrashed by various rightwing voices of the people.[29] Far from cultivating it, he merely capitalized on the uncomprehending rage that gripped the nation.

According to liberal columnist Joe Klein, to even consider that the US invited 9/11 with its merciless and single-minded foreign policy is "morally bankrupt."[30] It's fine to point to the mote in your brother's eye but recognizing the two-by-four lodged in your own is downright depraved.

The US couldn't possibly have invited 9/11, you see, because America has no interest in world affairs except to disseminate freedom and democracy, which we invented on July 4, 1776. Yet today these are little more than code words for capitalism. Democracy is the freedom of the moneyed class to buy politicians. Capitalism is a new religion, the blind faith in the unerring "invisible hand" of the market that provides for all our needs. The new sacrificial offerings are the hungry and sick, the jobless, the campesinos uprooted by "free trade" and exploited in maquiladoras. The economy giveth and the economy taketh away. It's no accident bin Laden targeted not only the Pentagon but the World Trade Center, chief symbol of America's permanent war against humanity and the earth.

Without the cover of Islamophobia, Obama could never have gotten away with killing bin Laden, a criminal who could easily have been captured and put on trial. Same goes for the murder of Muammar Qadhafi. Only in the context of the war pathology could Obama feel free to bomb Libya and arm its indigenous opposition, blocking negotiations and a proposed ceasefire, simply for geopolitical gain. Aside from imposing a steep toll on oil companies operating in Libya, for many years Qadhafi had been chief proponent and largest financial backer of African unity, a prospect enormously threatening to Western extractive interests.[31] That Qadhafi was yet another megalomaniac in no way justified intervening in the affairs of Libya, not to mention hunting him down and summarily executing him. Yet the only scandal to result from all this, promoted mostly on Fox News, was the subsequent attack on the US consulate in Benghazi. As far as the establishment goes, the left is rational-crazy while the right is crazy-crazy.

We can't seem to get a handle on war because we see only its explications — tanks and soldiers and missiles — overlooking the psychology that infects us from inside. Plenty of apparently rational motivations can be invoked for going to war, but without the malign influence of General Insanity and his sidekick Major Idiot, the idea of solving problems through systematic killing on a mass scale would be unthinkable. War has its own life apart from human reason. Sprung from the great unconscious, the presence of humankind past, Moloch self-propagates via dread oozing up unprocessed from a long buried sensation of helplessness. Our usual goodwill common sense shrivels in the face of a power so intense few can look it in the eye.

Like Lee Smolin says, it happens because it happened. With each new civilization arising from the ashes of its predecessor, the whole system reconstitutes itself like magic. But it's not magic.

It's morphic.

Still offering sacrifices to the beast lurking inside, still recapping our triumph over the jungle predator of old, humankind will remain in thrall to the family pet so long as we keep feeding it our blood.

1. Solomon, Norman, "The Media, Gulf War II & The FCC," *Z Magazine*, June 2003, p 3
2. Rampton, Sheldon, and Stauber, John, *Weapons of Mass Deception*, Tarcher/Penguin, 2003, pp 172,169
3. Hedges, Chris, "The Day That TV News Died," *Truthdig*, March 24, 2013. Available from: http://www.truthdig.com/report/page2/the_day_that_tv_news_died_20130324/

(Accessed Jan 3 2014)
4. Elliot, Michael, "America Stretched Thin," *Time*, September 1, 2003, pp 30-35
5. Shalom, Stephen, "Iraq: War and Democracy," *Z Magazine*, April 2003, p 4
6. Yoder, Steven, "Weapons of Mass Destruction, A Glossary of Terms", *Z Magazine*, October 2003, p 5
7. "Harper's Index," *Harper's Magazine*, October 2003, p 13
8. Anderson, Bruce, "Off the Record," *Anderson Valley Advertiser*, November 13, 2013, p 6
9. Blum, William, *Rogue State: A Guide to the World's Only Superpower*, Common Courage Press, 2000, p 29.
10. Barsamian, David, "Collateral Damage," *Z Magazine*, July/August 2003, p 50
11. Quoted in Ehrenreich, Barbara, *Blood Rites: Origins and History of the Passions of War*, Metropolitan/Henry Holt, 1997, p 77
12. *Ibid*, p 40
13. *Ibid*, p 39
14. *Ibid*, p 122
15. *Ibid*, p 123
16. Quoted in Ehrenreich, p 124
17. *Ibid*, pp 30-32,
18. *Ibid*, pp 73-74
19. *Ibid*, pp 70-72
20. "What Are You So Scared of? Saber-Toothed Cats, Snakes, and Carnivorous Kangaroos," Dunn, Rob, *Slate*, October 15, 2012. Available from: http://www.slate.com/articles/health_and_science/human_evolution/2012/10/evolution_of_anxiety_humans_were_prey_for_predators_such_as_hyenas_snakes.single.html (Accessed Jan 3, 2014)
21. Ehrenreich, pp 51-52, 87
22. *Ibid*, p 64
23. *Ibid*, p 63

24. *Ibid*, p 114
25. Sardar, Ziauddin, and Davies, Merryl Wyn, "Why Do People Hate America?", *Disinformation*, 2002, p 146
26. *Ibid*, pp 146-149
27. *Ibid*, pp 153, 180
28. Moyers, Bill, "LBJ's Path to War," *PBS*, November 20th, 2009. Available from: http://www.pbs.org/moyers/journal/1120 2009/profile.html (Accessed Jan 3, 2014)
29. Sardar and Davies, pp 182-185
30. *Ibid*, pp 35-36
31. Glazebrook, Dan, "The West's War Against African Development Continues," *counterpunch.org*, February 15, 2013. Available from: http://www.counterpunch.org/2013/ 02/15/the-wests-war-against-african-development-continues/ (Accessed Jan 3, 2014)

Chapter 13

Antisocialism

According to the average economist, the free market is the ideal mechanism for routing products to consumers, not to mention capital to companies and workers to jobs. So long as participants "maximize their utility," the market is unerring in its allocation of goods, wealth and talent.

Pricing on the basis of supply and demand can indeed simplify economic decisions, though not nearly as much as market ideology simplifies life for economists. According to business journalist Robert Kuttner, the "Platonic ideal of a pure market" frees theorists from pesky topics like history and politics, otherwise known as the real world.[1]

Like biologists reducing life to mechanisms or physicists rendering the flow of existence into static equations, market-fixated theorists replace reality with Quantopia and describe that instead. As Kuttner says, economists need only "devise the models, collect the statistics, and crunch the numbers."[2] Overlooked is the tendency of capitalist markets to concentrate wealth, leaving millions impoverished without hope.

Author of *Everything for Sale*, Kuttner explains how economists get lost in a thicket of quantitative abstractions. "A perfect market can be modeled; the analyst can perform neat simulations, using very sophisticated mathematics."[3] The "stylized algebra" of economic discourse omits corporate power over markets because this factor can't be modeled as easily as marginal adjustments in prices.[4] Whether in natural science or economics, mathematical modeling brings with it a degree of prestige that can't be attained through the mere gathering of real-world evidence.

Perhaps the biggest selling point of the pseudoscientific

laissez-faire model is its "congeniality to America's biggest businesses."[5] It pays to say what power wants to hear.

To the extent that it functions as a cult of the market, the economics profession views people as walking egos, our sole concern the acquisition of whatever is personally beneficial. Since society, by this view, is only an agglomeration of private spheres of interest, the public sphere has no function expect to interfere with the quest for self-gratification. Thus it's perfectly wholesome for a corporation to target kids with commercials promoting the health benefits of sugar-coated cereal, while a teacher explaining nutrition in school is indoctrinating children held hostage by the state.

Like Christian fundamentalists whose faith means they can do no wrong, the blind pursuit of self-interest provides a "blanket moral exemption" in the words of liberal economist Robert Heilbroner.[6] Scraping off mountaintops to expose deposits of coal must be good since it benefits coal-mining companies and their customers. Lending money to the desperately poor at 30% interest must be good since the lender makes a big profit and the borrower voluntarily enters into the exchange. A history of exploitation leading up to the borrower's poverty need not be considered, as history is excluded in favor of timeless equations.

In the logic of Quantopia public sanitation ought to be neglected because no one makes a killing from that. Wages ought to plummet as jobs are auctioned off to the lowest bidder. Even profits should disappear as each company seeks to undersell the other. The aviation industry should never have gotten off the ground since no one saw any interest in providing capital for a market that didn't yet exist. Across industries, no one should innovate because competitors will copy new techniques and cancel out the gains for innovators. No one should trust anyone else, since a client or business partner will break a contract as soon as doing so is advantageous. As Kuttner observes, it's paradise alright but only if you're a lawyer.

The idea behind unfettered capitalism, that everyone benefits if we all just stop cooperating and act only in our own private interest, is perfectly inverse to reality and therefore insane. Removing public regulations consistently generates results radically at odds with projections, as with airlines, electricity and banking. Yet market economists faced with the disaster of deregulation invariably argue that it failed because it didn't go far enough.

Market theory is commonly at odds with flesh-and-blood reality. The most "efficient" approach to workplace safety, for instance, is the cost-benefit analysis, whereby the cost of safety standards is balanced against the benefit to workers. In order to quantify this human good, economists peg the value of workers to their earnings. If a worker who earns $30,000 per year is killed on the job, this is seen as inefficient if the cost of safety measures is only $20,000 per year. However, if it costs $40,000 to protect the worker, better to let him die. Underdeveloped countries are so attractive to investment in part because the lower their wages, the more disposable the workers.

Though imposing safety standards based on nonmarket valuation of human life is thought to be inefficient, their actual effect is to encourage competition around improvements in the production process. Society not only promotes the intrinsic value of human life but stimulates gains in workplace efficiency.[7]

In a world where all profit-making activity is by definition a social good, even a system that profits from sickness must be beneficent. In the real world, by contrast, doctors increase earnings by authorizing unnecessary treatments while private insurers maximize profits by denying care, triggering an epic battle of the bureaucracies, each side hiring squadrons of paper-pushers to attack the other. As a result, in the late twentieth century, as the US supply of doctors grew 60%, the number of medical administrators rose six-fold.[8] Socializing health insurance and extending coverage to the entire population would reduce overall healthcare expenditures, but this option is "off the table" because sickness would be less profitable.[9]

The true patient, you see, isn't the old lady with a broken hip or the infant struggling for every breath but *Mammon*.

Capital, as Marx pointed out, expands in value by flowing through the production process. Starting in the form of money invested in a factory, capital materializes as a product and then reverts to monetary form when the product is sold on the market.[10] The goal isn't production per se so much as profiting from it. The more money that courses through Mammon's veins, the more wealth and power that accrues to the already rich and powerful.

Marx suggested producing directly for use rather than for sale on the market, reconfiguring economic activity in the interest of human freedom and well-being.[11] Like biologists breezily dismissing the self-determination of organisms, economists typically brush aside cooperative economics as idealist fantasy, projecting their own quantitative utopia onto those who would lead us back to reality.

As the stupendous productivity of industry saturates markets, capitalists are driven to find alternative investment outlets for their excessive profits. Rather than invest in the elimination of poverty, an unacceptable "utopian" act, the bulk of this wealth goes into speculation, whether in currencies, real estate, commodity futures, etc. Since heavily investing in a particular economic sector inflates its monetary value, the typical outcome is a bubble. The bursting of the housing bubble in October of 2008 literally bankrupted the US banking system.

When Mammon checked into the hospital, the doctors couldn't figure out how he could still be alive given that his heart had seized up and his cells were starved for credit. So Mammon's care-givers brought out the needle and gave him an "equity injection" on the order of $750 billion. Though it took many more such injections over the following years to bring the patient back to raging demonic health, at least he was no longer flatlining.

By market logic the banks that crashed the financial system should have been allowed to fail. This logic is especially compelling since "free market" banking deregulation triggered the crisis in the first place. According to Richard Fisher, president of the Federal Reserve Bank of Dallas, "institutions that amplified and prolonged the recent financial crisis remain a hindrance to full economic recovery and to the very ideal of American capitalism." Avoiding financial crises, he writes, requires the "fortitude to break up the giant banks."[12] But this solution is unthinkable to a Congress under the influence of big money.

Mammon loves market ideology as long as it justifies deregulation and obscene profits. But the moment crisis erupts, he's clamoring for government intervention, in this case bail-out of the very interests that drove the financial system off the cliff. And who are we to protest? Mass acquiescence to state-mandated robbery follows from mass dependence on the continued viability of that "world grasping commercial synthesis we call capitalism."[13] Without the transnational system of banks, manufacturers, shippers and merchants, where would we get our computers, our cars, our groceries?

In the early modern era, English peasants were systematically uprooted from commonly held lands and driven to poverty and crime in the cities. Known as the enclosure movement, this process has been recapitulated in the US via corporate takeover of agriculture under the motto, "Get big or get out." The loss of land means dependence on others for our livelihood. To have to work for an employer is to reduce oneself to the status of virtual slavery. But the memory of our collective loss of freedom has been buried, relegated to the collective unconscious, while we fixate on shopping, entertainment, spectacles, etc. Instead of acting to restore autonomy and freedom, we substitute self-determination with self-advancement, perpetuating the system that disenfranchised us in the first place.

Ironically it was precisely the attempt to avoid the dependency trap that triggered America's aptly named War of Independence.

In 1773 the East India Company hit a snag. Known around the world as simply "the Company," it boasted its own justice system, including jails, and an army of 250,000 soldiers, twice that of England's. It was so powerful that when a Parliamentary committee accused it of banning local trade and industry so as to create captive markets while extracting revenues so voraciously as to leave whole populations destitute, the Company wasn't the least bit concerned. After all, back in 1709 when Parliament revoked its charter, the Company brazenly ignored the order to disband, leaving competitors no choice but to merge with it.[14]

Then as now, leviathans need lots of income to stay afloat. The onset of European depression in 1772 posed a real dilemma, as the Company couldn't sell all that tea it was importing from its plantations in India. So its managers came up with a plan: the excess tea would be sold in the American colonies below cost. The Company would more than offset the loss by allowing only its own outlets to market the tea, bypassing established trading houses. Americans widely feared that once the Company drove tea merchants out of business, it would do the same to traders in other goods.

As a Pennsylvania pamphleteer put it, "if they succeed, they will send their own Factors and Creatures, establish Houses amongst us. Ship us East-India goods … till they monopolize the whole Trade. Thus our Merchants are ruined, Ship Building ceases. They will then sell Goods at any exorbitant price."[15]

The result was the Boston Tea Party. This was no symbolic display of outrage but a calculated effort to destroy enough tea to derail the Company's plans. The 90,000 pounds of tea dumped in Boston Harbor comprised roughly one-third of a year's supply. The empire struck back by closing Boston Harbor and banning the Massachusetts Assembly, which only inflamed the opposition and drew the previously fractious colonies into coherence. War was only a matter of time.[16]

A modern day equivalent of the Boston Tea Party would be labeled an act of terrorism, its participants hustled off to prison where they might or might not enjoy their right of habeas corpus. At some point between the 1770s and the present, Mammon got a foothold in America and grew up among us as one of our own. Unable to distinguish the social body from the infection, our national immune system attacks its host on behalf of the invader.

James Madison and Ben Franklin both tried to include restraints on corporate power in the Constitution, but each was rebuffed on the grounds of federal infringement on states' rights. In 1816 Thomas Jefferson expressed his hope that America would "crush in its birth the aristocracy of our monied corporations," but the states proved unable to take on Mammon. As steamboat pioneer Robert Fulton put it, the East India and Guinea companies "blindly extirpate one half of the human race to enrich the other." Yet Fulton, under Franklin's encouragement, obtained a corporate charter to build canals.[17] The lure of the corporate form was that it facilitated large-scale projects by pooling funds from numerous investors. Of course, such projects could also be carried out by the states and funded through taxation, but then who would profit besides the public?

When Pennsylvania Railroad was privatized, the legislature included a number of safeguards in its charter to prevent it from accumulating power. The corporation was to be governed by a yearly town meeting complete with public access to all its records. It could own only the land it needed for business and had to prove its dedication to the public good every 20 years when charter renewal came up. Just in case things got out of hand, Pennsylvania retained the right to buy it back at any time.[18]

Yet all this was for naught. By the aftermath of the Civil War, Pennsylvania Railroad had broken its shackles to emerge as a corporate juggernaut. Through bribery and deception, it got whatever it wanted everywhere from townships to the statehouse. The end result was not just rail service but concentration of power among private interests, corrupting the legislature and generating vast fortunes to finance predatory expansion into other states.[19]

By the turn of the twentieth century, corporate charters no longer included restrictions or threats of revocation. The modern corporation is immortal, can conduct business outside the bounds of its charter and can own other corporations. This last benefit turned out to be especially important, as it allowed a corporation operating in a more restrictive state to sell itself to a "holding company" in a less restrictive state, setting off a race to the bottom as states competed to be the most accommodating to their new corporate overlords. Between 1898 and 1904, a total of 1800 firms consolidated into 157 goliaths that dominated markets far and wide.[20]

A market under the control of a corporate cartel is not a true market, since the cartel can set prices regardless of supply and demand. The unspoken agreement is that prices are kept high enough to ensure profits. Since they don't compete with each other on the basis of price, corporations are left with advertising as the primary mode of competition in pseudo-markets. Commercials serve the dual function of promoting the dominant players and giving the impression of genuine competition.

Just as corporations seek to monopolize markets, the scientific establishment seeks to monopolize knowledge of nature. Fixated on prestige and authority, science never notices the schism between biology as cult of the molecule and physics as cult of the equation. Economics takes it a step further by establishing fundamental contradiction within a single field of study. By seeking to remove restrictions on corporate behavior, the cult of the free market encourages monopoly power. In most markets the only real freedom is that of the powerful to dominate them, including controlling what people know about the products they consume. It was precisely to prevent the dissemination of accurate information on products — the basis of free choice — that prompted Monsanto to pour millions into defeating California's ballot proposition to require labeling of genetically modified foods.

Though clearly at odds with genuine market principles, the corporate system was selected in the marketplace of possible economies, an outcome made possible by extending the market into the political arena with the wholesale purchase of legislators and judges.

Beyond the reach of voters, the Supreme Court is well placed to provide the legal basis for corporate empowerment over the public. The opening salvo came in 1819 when Chief Justice John Marshall struck down New Hampshire's attempt to turn Dartmouth College into a public school, arguing that the state was bound by Dartmouth's corporate charter, never mind that it had been signed by none other than King George III. The unstated implication was that Dartmouth enjoyed human rights, specifically the right to endure as a private entity.

Subsequent judges explicitly redefined the corporation into a kind of person with constitutional protections. Originally the concept of corporate personhood referred only to the rights to own land, enter into contracts and seek redress in the courts. In 1877 Supreme Court Justice Stephen Field, a friend of big-money interests, argued that all rights granted to human people also apply to corporate "people." Field was referring in particular to a phrase in the 14th amendment: "nor shall any State deprive any person of life, liberty, or property, without due process of law."[21]

It was Field who twisted a guarantee to former slaves into a rationale for granting corporations the liberty to pursue profit regardless of social consequences.[22] Though his colleagues were unconvinced, Field's warped take on the 14th amendment eventually won over a majority of fellow justices, and the workplace began to feel the effect in 1905. In the Lochner decision of that year, the Supreme Court struck down a New York law that limited the working hours of bakers to 60 per week, claiming the restriction violated due process for corporations.[23] By the mid 30s, scores of statutes had been invalidated on this basis, including child labor prohibitions, workplace safety standards, compensation for on-the-job injuries and union led boycotts of offensive employers.[24]

Finally, in 1937, popular pressure brought the Supremes to their senses, and public restrictions on corporate misbehavior were no longer struck down according to the 14th amendment. But the personhood model had taken root, and the Court turned to the Bill of Rights to shield corporations from federal oversight. Under the 1st amendment, they were granted the right of free speech, conveniently including the right to lie and to withhold relevant information from the public. The 4th, 5th and 7th amendments have similarly been misapplied to cases involving predatory corporate behavior.[25]

The point of constitutional rights is to shield people from concentrated power. Yet the courts call powerful corporations people in order to shield them from democratic oversight. That a corporation is obviously not a person, any more than it's a basketball or a lima bean, has no effect on judicial rulings. Not only are judges still vulnerable to corporate influence over their decisions, but more fundamentally the delusion of corporate personhood has carved a groove in the elite psyche.

Restrained by the conscience of their owners, ordinary businesses typically operate with a modicum of humanity. Not so for large-scale corporations. Shareholders have virtually no impact on the operation of the firm they ostensibly own. Meanwhile the CEO and other top executives are paid to focus exclusively on profits. The result is a schizoid separation of conscience from will. Those who might take moral responsibility for the actions of a corporation are out of the loop while executives systematically externalize costs onto employees, consumers and environments. If predatory corporations resemble a personality type, it's the *sociopath.*

Like the human race itself, madness has evolved. Simplicity has given way to subtlety and complexity, from brute psychosis to borderline condition to later, more refined ego disorders such as the attention-seeking histrionic, the compulsively controlling obsessive personality and the admiration-seeking narcissist, each stage a more refined form of schizoia. Where the narcissist neutralizes conscience with delusions of innate moral supremacy, in the sociopath conscience is excised completely. According to the American Psychiatric Association, *antisocial personality disorder* is diagnosed when three or more criteria are met:

1) *failure to conform to social norms with respect to lawful behaviors as indicated by repeatedly performing acts that are grounds for arrest*

Where the law conflicts with profit, corporations typically either flaunt it or pressure politicians or judges to strike it down.

2) *deceitfulness, as indicated by repeated lying, use of aliases, or conning others for personal profit or pleasure*

Aside from regular blasts of misinformation, whether aimed at the public or regulatory agencies, the corporation is itself an alias. Shareholders hide behind "Goldman Sachs" or "ExxonMobil" in order to avoid personal responsibility for theft and destruction.

3) *impulsivity or failure to plan ahead*

Did the big banks worry about the housing bubble they created? Did BP worry about poisoning the Gulf of Mexico when it successfully lobbied the Obama administration to ease safety restrictions? Does Big Ag worry about depleting soil and aquifers?

4) *irritability and aggressiveness, as indicated by repeated physical fights or assaults*

Corporate assaults can take the form of hostile takeovers of competitors or pre-emptive lawsuits to silence social or environmental critics.

5) *reckless disregard for safety of self or others*

This includes not only victims of toxic pollution and defective products but often the blue collar workers employed by the corporation itself.

6) *consistent irresponsibility, as indicated by repeated failure to sustain consistent work behavior or honor financial obligations*

Whether by underpaying workers to the point where they need food stamps to survive or snagging huge government contracts and then supplying a cheap product in return, corporations don't care. As to obligations to the communities that nourished them, they're happy to cut and run when a better deal comes along.

7) *lack of remorse, as indicated by being indifferent to or rationalizing having hurt, mistreated, or stolen from another*

The golden rule of the corporation is never to admit wrongdoing.

Like psychological regressives imposing unprocessed pain onto others, predatory corporations chronically externalize costs onto society at large, forcing the public to pay for environmental cleanup and provide welfare for underpaid workers. Setting aside CEOs, many of whom are indeed sociopathic, most corporate employees are considerate and caring people committed to doing the best work they can.[26] It's the corporation as a whole that operates like a psychopath, and it does so because it's been set up by law that way. As corporate executive Robert Monks says, "the enterprise has within it, and the shark has within it, those characteristics that enable it to do that for which it was designed."[27]

The corporation, says author and entrepreneur Ted Nace, is "programmed to survive, to maintain its structure and functional integrity, to grow, to avoid danger and recover from damage, to adapt, and to respond to the outside world." Like a shark seeking prey, there's "nothing malicious or even conscious about the tendency of the corporation to seek power."[28] It flexes political muscle to influence legislation over its legal framework "like a computer program capable of rewriting its own code." To understand the corporate threat, we ought to "pretend we are looking at a new form of life, a previously unknown organism."[29]

Or maybe we don't have to pretend.

The properties of the corporation built up over time not just in the law but in the implicate order. Just as a shark both expresses and reinforces the living form of its species, a corporation expresses and reinforces the forms implicit to its kind.

To repeat a behavior is to generate a habit, inscribing a pattern in the unconscious mind. When large numbers of people repeat a behavior, the resulting habit is ingrained collectively. The greater the number of past corporate executives who damage society and the earth for profit, the more likely that current executives will see nothing wrong with doing the same. Like embryos resonating with past similar embryos, corporate officers follow the developmental trail established by their forerunners. The only difference is that embryos rely on chemical interactions, while the corporate mindset arises from social interactions.

If Mammon represents labor in the service of elite wealth accumulation, the corporation is far from the only tool as its disposal. Class divide and exploitation date back to the early days of agriculture. After finally cutting off the head of the Company, nineteenth-century Britain serviced the beast via mostly standard business arrangements.[30] The corporate form has returned because it so reliably transmutes productivity for the people into power for Mammon. Contrary to the personhood model, what's corporealized is not human but demon.

As every person explicates the living memory of Homo sapiens, all large-scale profit-mad corporations explicate Mammon. But not only Mammon. Moloch too is implicit in their predatory operations. The twentieth-century emergence of the corporate economy recapitulated the division of Western civilization into power-hungry nation-states. Like the warring of early modern European powers, the never-ending battle for market supremacy ensures that whichever corporation prevails at any given time, the real winner is the ruling elite.

Systematic antisocialism amplifies the indifferent brutality of nature to the point where big chunks of humanity are so damaged and off balance that they can't fight back. But the war of the rich against the poor is only a side effect of the war of all against all. As JP Morgan Chase and other big banks demonstrated in the course of stealing billions from retailers via bogus credit card fees, corporations don't care whether their prey is human or corporate.[31]

So long as economic life is subsumed to the market mechanism, society remains atomized into competing interests seeking private gain at the expense of public good. Government is neutralized as a source of economic justice because money, by shielding the wealthy from hardship, decreases empathy even as it augments political influence. The citizen is bifurcated into employee and consumer, causing the first part to settle for an insecure unfulfilling job so the second part can buy cheap goods at a chain store. Contentment is discouraged both by fear of unemployment and the drive to get ahead. Meanwhile TV commercials tell us we're all happy and carefree. Periodic mass shootings aside, our ability to maintain a degree of sanity and compassion under the assault of institutional antisocialism only goes to show the depth and power of the human heart.

Marx thought capitalism would be buried under its own contradictions. Surely one of those cyclical busts will bring down the house. That hasn't happened in part because the system has a power over our minds, not just our pocketbooks. Having incorporated the beast, we focus on personal advancement without considering the globally destructive momentum to which we add our weight.

When Mammon contemplates itself by way of the economics profession, what does it find? Power prefers to explain the economy as a kind of algorithm gone mad. Simply by compounding the self-interest of each "player," we arrive at a self-regulating and maximally beneficial distribution of resources and wealth. Sure, things go wrong here and there, but that's only because the algorithm doesn't always behave. Funny how it keeps bolting off to ravage the countryside.

Naturally power airbrushes itself from its picture-perfect market utopia. By reducing Mammon to the chaotic playing out of an algorithm, the monster disappears. We're left with no sense of the whole organism — its purpose-driven behavior, its resilience and adaptability as it seeks to extract wealth from labor — no sense of the memory pool implicit in its every act, of the program running in parallel in every corporate mind, regressive or mutant. Like any algorithm, the one that allegedly animates capitalism is only a marker of where the beast has been over its evolutionary course.

If we see Mammon at all, it's in the outer mechanisms of capital accumulation but never the inner drive to rationalize social evils. Mammon inhabits not only skyscrapers and strip malls but the mental commons. The beast manifests in the unthinkability of revoking the charters of corporate predators or downsizing Pentagon procurement or socializing health insurance or incorporating ecological costs into economic transactions or investing in ghettos instead of attacking them in the guise of a bogus "drug war."

Continual televised stimulation of cravings combined with wealth inequality makes for despair and violence, feeding both Big Pharma and the ever more lucrative injustice system. Add to this a bloated army and a for-profit health insurance system, each of which diverts at least half a trillion dollars annually from productive use, and it's obvious the US cannot remain competitive, much less globally dominant, over the long haul. Even for the ruling elite, antisocialism as public policy is insane. As with war the lunacy is collective and self-perpetuating.

Just as our bodies develop and operate on the basis of species-wide habit, our worldviews and values get drawn into cultural norms that become collectively ingrained over time. The drive to exploit and profit is integrated into our mnemic makeup, as much a part of us as our toes and belly buttons.

A living thing doesn't go down without a fight. Alas this applies to Moloch and Mammon as much as you and me.

1. Kuttner, Robert, *Everything for Sale*, Chicago: University of Chicago Press, 1996, p 6
2. *Ibid*, p 34
3. *Ibid*, p 6
4. *Ibid*, p 12
5. *Ibid*, p 277
6. Quoted in Kuttner, p 50

7. *Ibid*, pp 300-02
8. *Ibid*, p 147
9. *Ibid*, p 155
10. Laska, P.J., "Nightfall for AWOL: The Dimming of the Dream and the Search for an Alternative," *Left Curve*, No. 35 (2011) p 6
11. *Ibid*, p 7
12. Fisher, Richard W, *Federal Reserve Bank of Dallas 2011 Annual Report*, pp 1, 23. Available from: http://www.dallasfed.org/assets/documents/fed/annual/2011/ar11.pdf (Accessed Jan 4 2014)
13. Shaw, Bernard, *The Quintessence of Ibsenism*, New York: Brentano's, 1915, p xvi
14. Nace, Ted, *Gangs of America*, San Francisco: Berrett-Koehler, 2003, pp 25-28
15. Quoted in Nace, p 44
16. *Ibid*, pp 42-45
17. *Ibid*, pp 46-48
18. *Ibid*, p 58
19. *Ibid*, pp 61-64
20. Bakan, Joel, *The Corporation: The Pathological Pursuit of Power*, New York: Free Press, 2004, p 14
21. Nace, pp 90, 94
22. *Ibid*, p 89
23. *Ibid*, p 127
24. *Ibid*, p 130
25. *Ibid*, p 167
26. Monbiot, George, "The 1% are the very best destroyers of wealth the world has ever seen," *The Guardian*, November 7, 2011. Available from: http://www.guardian.co.uk/commentisfree/2011/nov/07/one-per-cent-wealth-destroyers (Accessed Jan 4, 2014)
27. Bakan, p 70
28. Nace, p 226

29. *Ibid*, p 225
30. Bakan, p 6
31. Whitney, Mike, "The Humiliation of JP Morgan: Cut to the Chase," CounterPunch, Vol 19, No 13, July, 2012, p 5

Chapter 14

Disorganism

Despite utilizing many times more energy to produce the same amount of food as traditional "organic" farming, the industrial system of agriculture still manages to drain aquifers and deplete soil forced to yield the same profit-making crops year after year. Armed with petrochemical fertilizers, Big Ag keeps the plants growing but at the cost of nitrogen run-off breeding dead zones in oceans already acidifying to the point where all species of shell-forming sea creatures are threatened with extermination.[1] Meanwhile warming and acidification are putting the squeeze on phytoplankton, the base of the ocean food chain and source of half the oxygen we breathe.[2] Tropical forests, which provide much of the other half, are getting burned down acre by acre to make way for farms and pastures. With the removal of huge amounts of biomass, rainwater flushes out what's left of biodiversity in the mad rush to river and sea, threatening to tip the entire Amazon basin into semi-arid savanna. The only question is when.

With every ecosystem on Earth visibly in decline, the rate of species extinction is already 100 times the pre-industrial background rate, making this the first mega die-off in 65 million years.[3]

Ecosystems typically disintegrate when 50-90% of their original expanse has been eradicated due to agriculture, mining, paving, etc. According to ecologist Anthony Barnosky, this applies globally as well. So we might be concerned that more than 40% of Earth's productive land is now appropriated for exploitation. Once most of the world's usable lands are under (in)human domination, it's only a matter of time before the planetary eco-tapestry that birthed our species unravels.[4]

Even as we slurp down 90 million barrels of fossilized life every day, we can't seem to avoid threatening the viability of millions of current species, including our own. How did we come to this?

The temptation is to blame overpopulation or pesticides or factories or gasoline, but the real enemy is the cancer of compulsive growth endemic to capitalism, "that ubiquitous, all-powerful and greatly misunderstood dynamo," says Joel Kovel, "that drives our society."[5]

Author of *The Enemy of Nature*, Kovel explains the fundamentals. While the traditional marketplace is a means of exchanging goods for money so as to purchase other goods, under capitalism the market mutates into a mechanism for accumulating wealth. Reversing the natural order, the merchant starts off with money and buys the product of someone else's labor, then turns around and sells that product at a markup. As long as the laborer is paid less than what the product is worth, the trader makes a profit.

Under capitalism what gives a commodity value is not what we do with it, like using bricks to build houses or shoes to walk home in, but the price it commands in trade. In contrast to "use value," a quality that belongs to an item intrinsically, "exchange value" is an abstraction that can only be expressed quantitatively. When you buy a pair of shoes — or better yet a thousand pairs — only to sell them for profit, their value is reduced to a number.

As the basis of economics becomes the trade itself and not the tangible thing exchanged, money is transformed into an all-consuming monster. No longer bound up with the limitations of actual land, people and resources, it springs to life, an abstraction with a will of its own. "Pure quantity," writes Kovel, "can swell infinitely without reference to the external world."[6]

There lies the source of our ecological crisis.

Despite its reputation as the acme of rational economic exchange, capitalism follows its own imperative apart from the needs of humans and ecosystems. In its compulsion to grow and multiply, capital seeks to violate whatever limit is set before it. Success means only one thing: surpassing yesterday's mark. No matter how big the corporation gets, to cease growing further is to die, to be overtaken and swallowed up by competitors. Yet the one thing we know for sure is that the system can't grow forever. The world contains only so many untapped resources and markets. Sooner or later abstraction runs up against reality.

The rational response is to adapt, to change behavior in the face of changing circumstances. By contrast, the capitalist response is to exploit more intensively, as in the case of hydraulic fracturing, a.k.a. fracking, whereby chemical pollutants are injected into the earth, often winding up in underground water supplies, so as to free up otherwise inaccessible pockets of natural gas.[7] To get our hands on more coal, we're blasting off mountaintops, burying rivers in debris and amplifying rates of terminal cancers and birth defects among people unfortunate enough to live nearby.[8] Instead of reducing dependence on fossil fuels, we engage in costly procedures, both monetary and environmental, to extract what's left as quickly as possible. It makes no sense – unless, that is, you inhabit a world made of numbers.

Already in the nineteenth century it was clear that gains in energy efficiency, from Watt's steam engine onward, nearly always lead to greater use of energy. This is known as the Jevons paradox after the economist who noticed this tendency.[9] What Jevons perhaps didn't realize is that capitalism wouldn't have it any other way. So long as money seeks outlets to generate more money, consumption rises to match whatever amount of energy is available. We can't scale down our use of resources, because our own economic system is sabotaging us at every turn.

As we saw in the case of the stratospheric ozone hole, some environmental threats can indeed be resolved. Scientists might find an alternative to the pesticides that appear to be wiping out bee populations necessary for much of our agricultural output.[10] As historian Arnold Toynbee pointed out, civilizations are made to overcome challenges. Yet time and time again cities have become ruins because a challenge came along that was too great. No matter how many bullets we manage to dodge, capitalism keeps putting us right back in the line of fire. Eventually a problem will emerge — or more likely a complex of interrelated problems — that won't yield to a technological fix.

Marx thought revolution was only a matter of time. Capitalism, he said, cannot help but prepare the stew in which it will roast. As long as profit requires underpayment of labor, working people are chronically short of the cash needed to purchase the products of industry. Yet we've seen the system respond to crises of overproduction in numerous ways, including the recent encouragement of mass indebtedness. So Kovel takes the next step. Instead of periodic economic downturns, the shattering of Earth's life-giving integrity is destined to set off socialist (make that ecosocialist) uprising.

What Kovel overlooks is the likelihood that worsening environmental conditions will exacerbate the scarcity that already pits us against each other. If the money-driven struggle of all-against-all is pushing us, inexorably, to the edge of the cliff, shouldn't we expect rising insecurity and the resulting intensification of this struggle to push us right over the edge?

Kovel himself points out that capitalists are perfectly willing to perpetuate eco-destabilization as long as they can insulate themselves and perhaps even profit from the meltdown around them. He cites an article in London's *Guardian Weekly* showing a shift in elite opinion since the early 70s, when the Club of Rome called for "limits to growth."[11] These days, digging our own grave is the ultimate business opportunity.

Monthly Review editor John Bellamy Foster nicely sums it up:

We should not underestimate capitalism's capacity to

> accumulate in the midst of the most blatant ecological destruction, to profit from environmental degradation ... and to continue to destroy the earth to the point of no return – both for human society and for most of the world's living species.[12]

Times are tough? How about a liquidation sale? Like Marx before him, Kovel finds a silver lining where none exists. While looming disaster may indeed spur much-needed green reform, especially in food and energy production and distribution, the system is perfectly capable of incorporating these changes and carrying on with its prime directive of endless growth. There's just no pulling the socialist rabbit from the capitalist hat.

Not so long ago Americans mostly made their living off the land with the freely offered assistance of neighbors. Today the few remaining family farms no longer rely on mutual aid so much as the costly assistance of "machines, chemicals and credit," as Wendell Berry puts it.[13] The new order, he writes, "is the product of an abstracting industrial and mercenary intelligence, alien both to the nature of the land and to the minds and lives of the people."[14]

In the context of global ecological meltdown, the old metaphors of Moloch and Mammon no longer suffice. Like two strands of DNA in coiled embrace, M&M have combined to give us a new and improved beast, a *disorganism* whose body is fashioned from computers, roads, refineries, malls, prisons and weapons while its mind is our own blind habits of dependency, conformity, careerism and lethal indifference toward the not-us. All of these characteristics, which built up over countless generations during prior civilizations, are now implicit in our behaviors, ingrained in the hive mind.

We all play a part in feeding disorganism, whether ordinary people trying to get by or the rich trying to get richer, though the ones benefiting the most from this arrangement bear far more responsibility. A few even play an active role in cultivating an environment conducive to its continued predations. As Robert Fitch demonstrates in *The Assassination of New York*, starting in the 1970s a clique of investors and city officials meticulously planned the replacement of local manufacture with finance, insurance and real estate, undercutting working families while augmenting the authority of concentrated wealth.[15] Though such behind-the-scenes planning is occasionally necessary to keep the beast on the prowl, mostly elites just let it follow its nose.

Disorganism is made of the same stuff as you and me. It lives in our resonance with like-minded predecessors from military-commercial empires of old. It manifests in our inability to conceive of any response to intensifying eco-crises except to go faster down the path already leading us to disaster. Having no idea it's upending ecosystems, it just wants to lay its eggs in as many locations as possible. It does good as blindly as evil, such as when speculators bid up the price of oil, inadvertently reducing usage, or bankers crash whole economies and thereby slow consumption. Meanwhile we internalize its logic, viewing the ruined farmer or the diseased miner as perfectly natural byproducts of "free enterprise." The individualist ideology we're so proud of merely cloaks our collective subordination to the beast.

Its source of power is mass dependence originating in loss of land and relative self-sufficiency. Britain's enclosure movement is echoed today in "free trade" agreements that force campesinos off the land when they can't compete with subsidized industrial agriculture.[16] Even in ostensibly communist China, the drive for capital accumulation is eliminating 10,000 small farms every day, feeding a twenty-first-century trail of tears into metastasizing cities.[17] Wherever they take place, mass displacements are steps in the dance with disorganism.

Rather than confront the beast, our society projects its own degradation onto "sinful" activities like prostitution, drug use and gambling. As long as we fixate on the alleged evil of the other, we don't notice that our whole way of life is built around the prostitution of our talents to big money and that we're lost in a haze of petrochemical addiction as we casually gamble with the future of life on Earth.

Disorganism skillfully adapts to increasing social and environmental instability as it steers us down the road to universal holocaust. When ordinary people can't afford the products of industry, the beast breaks left. When capitalists can't accumulate enough wealth for investment, it breaks right. By extending the era of maximized production and consumption to the very last possible moment — all the while eroding our ability to care for ourselves locally — disorganism intensifies the impact of collapse once it arrives, rendering not only humanity inhuman but life unlivable.

Like Mayans sacrificing their neighbors en masse to ensure the next sunrise, we've set aside actual needs to service the demands of an insatiable abstraction, even if this means destroying forests, exhausting soil, depleting and poisoning groundwater, acidifying oceans and tampering with everything from climates to DNA. Already three billion people lack decent housing, steady food supply, clean water and medical care, victims of a system that generates poverty as a toxic byproduct of wealth concentration. The meaning of ecological crisis is that we're also concentrating wealth in the present at the expense of the future. To see what we're ultimately doing to ourselves, we need only examine the deprivation imposed on others today.

Because each of us expresses the same pool of living memory, everyone who ever lived continues to live in us, and everyone to come is only us recycled. No sense grabbing all the goodies now when those who follow us are still us. By screwing the future, we're screwing ourselves. It's nuts, and the insanity is irreducible to individuals, since most of us are perfectly sane. We've mutated into individuations of disorganism as much as Homo sapiens, laboring for *its* propagation instead of our own.

In a little corner of New Guinea, a people called the Korowai have a strange habit of killing in order to prevent killings. When a sick person nearing death identifies another member of the tribe as the cause of his demise, that person is deemed a khakhua and must be killed and eaten so as to scare off other khakhua. Korawai insist they eat only khakhua, not humans. "It's normal," says Bailom. "I don't feel sad that I killed Bunop, even though he was a friend."[18]

The Korowai autodelusion is like a bacterium compared to the multicellular behemoth now raging across the earth, enslaving in the name of freedom, impoverishing in the name of prosperity and killing us all for the betterment of mankind.

How do we fend off this creature that grew up out of our own tendencies and even now sees through our eyes and speaks through our mouths?

The mutually reinforcing tendencies of mutancy and regression have morphed America into the Wicked Witch of the West. US leadership thinks it can manage the human race and the earth for private gain at the expense of living creation — *and no one will have any problem with this*. The human portion of the world's expendable biomass will join millions of animal and plant species in the noble endeavor of quietly disappearing, leaving only Quantopia, paradise for the tiny elite who engineered it all.

From Plato to this. The illusion of escape from open-ended presence to perfect immutable being has become autocide, the sloughing off of human deadweight in the midst of generalized eco-destruction, all of it set in motion mostly by mistake by people who think their temporary advantage over the other losers somehow makes them winners.

The capitalist elite have no idea they're servicing the needs of something inhuman, an implicate being with no concept of us or that our annihilation will spell its own demise as well. Only on occasion does a Quantopian, a master of the universe, reflect long enough and quietly enough to glimpse the magnitude of the crime underway.

1. Magdoff, Fred and Foster, John Bellamy, *What Every Environmentalist Needs to Know About Capitalism*, New York: Monthly Review Press, 2011, pp 15, 17, 20
2. Roach, John, "Source of Half Earth's Oxygen Gets Little Credit," *National Geographic News*, June 7, 2004. Available from: http://news.nationalgeographic.com/news/2004/06/0607_040607_phytoplankton.html (Accessed Jan 4 2014)
3. Magdoff and Foster, pp 21, 77, 18
4. Barnosky, Anthony D, et al, "Approaching a State Shift in Earth's Biosphere," *Nature* 486, June 7, 2012, pp 52-58
5. Kovel, Joel, *The Enemy of Nature: The End of Capitalism or the End of the World?* London: Zed Books, 2002, p 38
6. *Ibid*, p 128
7. Hightower, Jim, "What the Frack!?!" *Hightower Lowdown*, Vol. 14, No. 7, July 2012. Available from: http://www.hightowerlowdown.org/node/3039 (Accessed Jan 4 2014)
8. Howard, Jason, "Appalachia Turns on Itself," *The New York Times*, July 8, 2012. Available from: http://www.nytimes.com/2012/07/09/opinion/appalachia-turns-on-itself.html?_r=3& (Accessed Jan 4 2014)
9. Foster, John Bellamy and Clark, Brett, "The Planetary Emergency," *Monthly Review*, December 2012, p 6
10. Pesticide Action Network, "Beekeepers Ask EPA to Remove Pesticide Linked to Colony Collapse Disorder, Citing Leaked Agency Memo," December 8, 2010. Available from: http://www.panna.org/media-center/press-release/beekeepers-ask-epa-remove-pesticide-linked-colony-collapse-disorder-citin (Accessed Jan 4 2014)
11. Kovel, p 47
12. Foster, John Bellamy, "Capitalism and Ecology: The Nature

of the Contradiction," *Monthly Review*, September 2002, pp 10-11

13. Berry, Wendell, *Home Economics*, New York: North Point Press, 1987, p 174
14. Quoted in Magdoff and Foster, p 72
15. Henwood, Doug, LBO News, "Explaining what goes on in the world: in memory of Bob Fitch," *LBO News*, May 22, 2012. Available from: http://lbo-news.com/2012/05/22/explaining-what-exists-in-memory-of-bob-fitch/?utm_source=feedburner&utm_medium=email&utm_campaign=Feed%3A+LboNewsFromDougHenwood+%28LBO+News+from+Doug+Henwood%29 (Accessed Jan 4, 2014)
16. Public Citizen, "Unfair Trade" November 2003. Available from: http://www.nffc.net/Learn/Reports/unfairtradepc.pdf (Accessed Jan 4, 2014)
17. Magdoff, Fred, "21st Century Land Grabs: Accumulation by Agricultural Dispossession," *Monthly Review*, November 2013, p 12
18. Raffaele, Paul, "Sleeping with Cannibals," *Smithsonian*, September 2006. Available from: http://www.smithsonianmag.com/travel/cannibals.html (Accessed Jan 4 2014)

Chapter 15

Whirlpool

Downstream from Niagara Falls a whirlpool drains water through its revolving eye, a task it's been performing without pause since the great melt-off that began 12,000 years ago. So lifelike is this autonomous force of suction that it has its own name. *Whirlpool* even has a story to tell. The message emanating from its hypnotic swirl is both a warning of what happens to societies that defy nature and an illustration of what we'll have to do to get back on track.

Ecologist Eric D Schneider saw the writing on the rippling wall back in the 70s. In search of a theoretical framework by which to comprehend the ecosystem collapse he witnessed in the course of his fieldwork, Schneider ran across the emerging science of nonequilibrium thermodynamics. At the crossroads of physics, chemistry and biology, NET explains how a complex self-organized system, like a whirlpool or a kidney cell or a modern human economy, maintains itself despite being perpetually out of balance with its surrounding environment.[1]

By human logic complexity is what happens when you compound simplicity. Nature begs to differ. Far from being built up gradually from simpler elements, complex systems emerge all at once in totality. Our friend Whirlpool is in no way reducible either to mathematical laws or the algorithmic behavior of its molecular components. Instead it's a spontaneous self-creation that lives to channel energy. With only a slight stirring of the waters eons ago, it sprang into action, its energy derived from wiping out the pressure gradient that results from Niagara-sized sheets of water trying to squeeze through the opening at the bottom of its basin. By rapidly closing the gap between where the water is and where it "wants" to be, Whirlpool is energized with spiraling lines of force that flush out water 30 times faster than simply letting it drop straight through the hole.[2]

If nature has an idea, it's that gradients are bad. Wherever there's a gap, nature rushes in to close it. Indeed the very flow of time follows from nature's continual attempt to eliminate the gradient between absolute presence and oblivion. From this we get the fleeting moments of sensory existence.

When air becomes segregated between warm and cold pockets, excited molecules are drawn into cooler areas, thereby gradually reducing the gradient. Complex weather systems like tornados and hurricanes eliminate temperature gradients with much greater efficiency, bursting into existence like a bolt of lightning equalizing the electrical charge of ground and sky.[3]

The natural tendency to even dispersal is encapsulated in the concept of *entropy*. You'd think life, which specializes in concentrating energy, could never appear in a universe where entropy is the "rule of rules." But nature approves of anything that rapidly breaks down gradients. As Schneider and fellow author Dorion Sagan explain in their whirlwind tour of NET, *Into the Cool*, organisms are gradient-reducing chemical systems that evolved the twin capacities of storing energy and reproducing. Nature loves life so long as living things procure their energy by reducing concentrations of it in the environment.[4]

But when organisms themselves become concentrations of energy, they have a problem. Now nature wants to extract it from them, and this tendency is measured by the amount of waste they produce. As a tree harnesses increasing amounts of sun-derived energy during growth, it must expel increasing amounts of degraded energy, i.e. entropy, in the form of heat. Animals have more elaborate disposal methods because they generate greater concentrations of energy flow, while garbage and pollution manifest the immense energy flows in human economies.[5]

The solution to energy loss due to entropy is the evolution of ever more complex structures that improve the efficiency of both energy capture and waste export. But the increasingly delicate, fine-tuned systems that result are also increasingly out of balance with their environments and therefore require more energy to maintain their structure. For this reason the history of life on Earth, including human society, is characterized by long periods of gradually increasing complexity punctuated by sudden and devastating crashes.[6]

"Life," say Schneider and Sagan, is a "cycling system whose physical, material, and ultimately mundane purpose is to get rid of prior complexity."[7] The degree of complexity attained in the past determines the degree of waste the current system must evacuate to avoid getting backed up. If you can't process the waste from yesterday's developmental achievement, you must draw in more energy, setting up an even bigger disposal problem for tomorrow, and so on.[8]

Archaeologist Joseph Tainter argues that civilizations aren't brought down by invasion or environmental calamity so much as the weight of their own prior development. In Rome, the conquest of distant lands initially fed the treasury but turned into a drag over time as the bureaucracy of empire bloated. Like a pyramid scheme, new conquests paid for the maintenance of already conquered territories. Once the boundaries were set, largely under the reign of Augustus Caesar, the seed of collapse was planted. The moment the ink was dry on the map, Augustus had to levy new taxes to make up for shortfalls in revenue. Over the following centuries, the continent-spanning infrastructure of cities, roads, ports and aqueducts, which had set off waves of prosperity as it took shape, became dilapidated and increasingly expensive to maintain. The tax burden on peasants got so severe as to drive many into bankruptcy and slavery. With the loss of countless farms, the tax base shriveled, necessitating higher rates on those that survived. A self-perpetuating cycle was unleashed that ravaged the empire far more effectively than barbarian armies ever could.[9]

The greatest strength of capitalism, according to Marx, is also its fatal flaw. So efficient is this complex system at generating wealth, investors eventually run out of profitable outlets for it, producing a kind of economic constipation and periodic crashes. As in the thermodynamics of life, yesterday's achievement becomes a burden that must be overcome by today's gains. "If capital increases from 100 to 1000, then 1000 is now the point of departure from which the increase has to begin ... What appeared as surplus-value now appears as simple presupposition."[10]

Just as energy flow generates complex organic forms, circulation of money stimulates production and distribution of goods. And just as chloroplast obtains energy for plants by tapping into the temperature gradient between sunlight and dark space, merchants derive wealth by exploiting the gap between what people want and what they have. The invisible hand of the market is nature's own. Capitalism works because it's natural.[11]

Or is it? Sure, you want energy flow but not so much that you drown in the waste accumulating in its wake. As they mature, ecosystems become populated chiefly by slow-growth plants, which extend lifespan by using energy more minimally and efficiently than the rapidly growing species they displace.[12] Forever stuck in the fast-paced juvenile phase of development, capitalism is unable to switch gears to a sustainable rate of energy harvesting and waste processing.

Under the reign of capital, we're never more than a step ahead of being swamped by the entropy accruing from prior overdevelopment. Locked in a hell of our own making, we must constantly accelerate money flow and maximize development, be it through conquest and colonization, intensified exploitation of labor, technological advancement or increasingly exotic financial instruments such as subprime mortgages and collateralized debt obligations. No matter how close the system takes us to meltdown, it always wants to push us that much further.

What makes capitalism self-perpetuating is its bipolarity of upper and lower class. Just as Whirlpool feeds off the pressure gradient resulting from a large body of water trying to fit through a small opening, capitalism feeds off the gradient caused by a large body of impoverished people trying to live a decent life. The key to capitalism is exploitation, the process whereby those who already have lots of money make even more at the expense of those with virtually none.

Modern poverty, as social critic Andre Gunder Frank points out, is not a leftover of some mythically all-impoverished past but is actively created in the course of capitalist development. Upper class and lower class are held at bay via the energy unleashed by tapping into the gradient between them. As Frank puts it, the "development-underdevelopment contradiction ... is the source of its development throughout."[13] Exploiting the poor prolongs their poverty and therefore the wealth gap that powers the system. In order to maintain the gradient off which it feeds, the system must simply continue feeding off it.

So long as Niagara falls, Whirlpool makes room for more. The energy that keeps it spinning is no different, fundamentally, than the self-perpetuating flow of resources from poor to rich. Throughout the global South, wealth is continually vacuumed up, whether via mining and monoculture, cheap labor for foreign-owned factories or interest on endlessly revolving debt. As surely as Whirlpool reproduces the conditions required for its survival, the money-driven economy perpetuates the wealth gradient.

In short, capitalism sucks. The alternative, by definition, is socialism. Rather than dividing society into one part that benefits at the expense of the other, a socialist economy operates for the benefit of society as a whole. Socialism can take many possible forms, from the workable to the disastrous, from organic to mechanistic, from anarchistic to authoritarian.

The worst kind of socialism is what Lenin and Stalin imposed on Russia after 1917. Communism seeks to equalize wealth by way of rapid, centrally planned development. Though devoid of the life-or-death competition that drives centralization in the capitalist world, communism arrives at the same end by transforming the state into a single mega-corporation that makes all economic decisions. The state corporation hires the entire workforce, grows all the food, builds all the housing and manufactures all the products its consumer-citizens buy at its stores. Communist logic is impeccable. Unfortunately, it's our logic, not nature's.

As the people of the Soviet Union learned, nature isn't shy about humbling those who concentrate too much energy and build up too much infrastructure all at once. All that development starts falling apart at the same time, and the system is crushed under its own weight.

Ultimately the West faces the same outcome. With its totalitarian system of command and planning, every corporation is a USSR in miniature. Regardless of whether we call it communist or capitalist, nature's response to centralization of power and wealth is to dine on the bloated flesh of the ungulate. Overseas exploitation can extend the autodestruct sequence, but this only makes the collapse that much more horrific and all-consuming once it arrives.

Following the Soviet collapse, activist and author Michael Albert attempted to take the Stalin out of Stalinism by proposing a system known as participatory economics, or *parecon*, which still relies on centralized production but utilizes the internet to allow ordinary people to vote on what they'd like the state to produce and at what price.[14] The problem is that no matter how democratic or participatory such a system is, it remains thermodynamically incorrect. Though warped under the weight of monopoly capitalism, the market is still a vastly more efficient way of determining production and pricing than planning every economic output down to the last detail.

Let's say warm air has congregated at ground-level under a heavy blanket of cold air. Nature is unhappy. Now let's say all that warm air is abruptly funneled into the sky, equalizing the temperature almost instantaneously. Without any planning, preparation or even announcement, a tornado has popped into being and gotten the job done. Nature is pleased. But if it was up to Albert, that tornado would have been assembled one speck of dust at a time, the location, direction and speed of each particle to be determined by popular vote. Even if it ever got off the ground, such colossal inefficiency couldn't stir up the frosting on a cake.

Like Plato, Albert presents a timeless vision of the good society but provides no realistic means of getting there or how to sustain it once we did. Perhaps we should take the lead of mathematician David Schweickart, who champions market socialism. According to this model, the public owns the means of production, but workers make the decisions of what to make and how to make it. Nobody can buy their way into ownership of a company because none of them are for sale. As Schweickart puts it, market socialism "maintains a competitive market for goods and services, but replaces the labor market with workplace democracy and the capital market with social control of investment."[15]

Because workers don't currently control the means of production, we must compete to sell our labor to those who do. By eliminating this market but keeping the one for products, Schweickart would banish exploitation while retaining market efficiency in production and allocation. No more exploitation means no more wealth gradient and no more impetus for unlimited growth. The economy, at last, would be allowed to mature into slow-growth long-term sustainability.

Or would it? From a thermodynamic point of view, the basic problem remains. Though worker-controlled, the revamped system would still be over-centralized, requiring so much energy flow that the resulting waste couldn't be exported without further growth and even greater energy flow. We'd still be on the same cycle taking us to the same old scrap heap of failed civilizations.

Both Albert and Schweickart depend on the emergence of a political movement potent enough to outlaw concentrated wealth and socialize large-scale private firms in manufacturing and services, an outcome inconceivable in the US. That's not to deny the importance of political movements, just that while forcing change through state mechanisms of election, legislation and lawsuit — though potentially beneficial in myriad ways — isn't likely to unravel a system on which all of us are hopelessly dependent for our material well-being.

Locked into growth and complexification, our economy is immune to calls for change. Like sick and wounded creatures cranking up the energy flow just before death, the system seems fated to feverishly cycle its way to final collapse.

We need a housekeeping system that not only operates but emerges in thermodynamic accord with nature. It starts with a stirring of the waters, so to speak, which sets in motion a system of production and exchange that operates on the basis of a new energy-producing gradient. Already incipient in the form of producer and consumer cooperatives, intentional communities, urban homesteads and alternative currencies, the social economy has the potential to exploit the inefficiency of a system which, in the process of concentrating power, renders masses of intelligent people into thoughtless consumers and laborers.

Just as capitalism generates the wealth gradient that fuels its cancerous expansion, the social economy produces an efficiency gradient between itself and the machinery of the dominant system. In contrast to the capitalist compulsion to increase production and consumption without limit, the social economy is based on judicious use of resources to maintain a modest standard of living. Decentralized organization serving the needs of society is inherently more efficient than top heavy organization serving to maintain its own power.

Alongside overseas investment and rising public and private debt, pointless spending is a prime means for continued expansion of capital in the face of markets saturated by mass merchandise. This is why corporate America blows more than a trillion dollars a year in marketing alone. The more money that goes toward marketing, packaging, luxury, production of items built to fail, transportation of food and other items over great distances, prisons for the superfluous poor at home and weapons to terrorize potential revolutionaries abroad, the greater the number of outlets for continued profitable investment but also the greater the waste inherent to the system as a whole. Much of our overconsumption results from dependency on employment – in order to produce all that capital-expanding waste – and the concomitant loss of time needed to care for our own needs. A social economic alliance would greatly reduce waste in the course of caring for itself, distributing goods freely among its worker participants.

To the extent that we produce our own goods, along the way unifying production and management into an organic whole, we gain efficiency relative to the system at large. But only insofar as we trade with the corpulent system – selling surplus and buying what we don't make ourselves – can we cash in on that efficiency gap, earning the money that fuels our metabolic growth. As the social organism propagates and extracts increasing amounts of energy via the efficiency gradient, the wealth gradient dries up, killing off capitalism at the source.

The idea of a self-propelled organic socialism has been in the air for a long time. Something like it was proposed by Marx's contemporary, Pierre-Joseph Proudhon, who called it *mutualism*.[16] So why, Schweickart wonders, hasn't the social economy popped into being by now?

Part of the problem is the tendency to lose steam and collapse back into the system. Quebec's attempt to secede from Canada in the 1970s sparked not only the usual demonstrations and occupations but self-governing neighborhood committees and the proliferation of worker-owned cooperatives. Though some of the cooperatives still operate, many more have since been "engulfed by the state," in the words of researcher Anna Kruzynski.[17] So too, many of the organic food production companies that fueled the rise of consumer cooperatives in the US have sold out to corporate interests.

Nonetheless at least two successful self-funded alternative economies have emerged in the last few decades: the Mondragon network of largely manufacturing cooperatives in the Basque region of Spain and the Landless Rural Workers Movement in Brazil. While both organizations are intended to benefit the regions where they originated – and neither adequately combines food and durable goods production – they reveal the potential for a universal social economy.

Complex chemical systems had to evolve to a high degree of efficiency before birthing free-living bacteria. We shouldn't be surprised if a social economy requires extensive development before it becomes self-propagating. A great deal of energy and wealth will have to be harnessed to give it form, to draw together radical currents into a coherent metabolic and reproductive system capable of converting its efficiency edge into sustained growth.

So long as wealth is concentrated and most people depend on the labor market for survival, we have no choice but to advocate for public regulation of private enterprise, for general welfare and government investment in job-creating infrastructure development. This puts the political left in a bind, since big stimulus today means big entropy tomorrow and the resulting drag on the economy. Every gain for the left sets up an opportunity for the right to slim down the system by slashing wages and social spending. By contrast, when a satisfactory standard of living is attainable through skilled labor in the interests of local community, neither market nor government need be sources of tyranny, and the economic dimension of the left-right divide is superseded.

The key is to stop thinking in terms of simple dichotomies – markets vs. socialism, private vs. public, daily life vs. political activism – and start thinking like nature. Revolution doesn't start from abstraction and work its way down to the proletariat. It's the culmination of a dialectic whereby economic development enables greater political muscle, which opens the door to further development, and so on.

This is precisely the dialectic that brought the capitalist whirlpool into being, and this is how it will be drained. Screaming over the swirling roar to demand justice is no better than dropping a bomb or two on it. Until we shut off the water supply, it will keep on spinning.

1. Schneider, Eric D, and Sagan, Dorion, *Into the Cool: Energy Flow, Thermodynamics and Life,* Chicago: University of Chicago Press, 2005, pp. 1-4
2. *Ibid*, pp. 111-116
3. *Ibid*, p. 6
4. *Ibid*, pp. 145-146
5. *Ibid*, pp. 43-45, 208
6. *Ibid*, p. 210
7. *Ibid*, p. 323
8. *Ibid*, p. 328
9. Tainter, Joseph, *The Collapse of Complex Societies*, Cambridge: Cambridge University Press, 1988, pp. 128-150
10. Meszaros, Istvan., *Beyond Capital*, New York: Monthly Review Press, 1995, p. 521
11. Schneider and Sagan, p. 276
12. *Ibid*, p. 158
13. Frank, Andre Gunder, *On Capitalist Underdevelopment,*

Oxford: Oxford University Press, 1975, p. 44

14. Albert, Michael, *Parecon: Life after Capitalism*, London: Verso, 2003, pp. 9-12
15. Schweickart, David, "Nonsense on Stilts: Michael Albert's Parecon," *Loyola University Chicago*, January 16, 2006. Available from: http://homepages.luc.edu/~dschwei/parecon.htm (Accessed Jan 4 2014)
16. Gambone, L., "Proudhon and Anarchism: Proudhon's Libertarian Thought and the Anarchist Movement" *Spunk Library*, 1996. Available from: http://www.spunk.org/texts/writers/proudhon/sp001863.html (Accessed Jan 4 2014)
17. Gottinger, Paul, "A Look Back at Maple Spring: An Interview with Anna Kruzynski," *Z Magazine*, July/August 2013, pp 22-23

Chapter 16

Fate

Imagine a world that makes sense.

It's the fall of 1971 and I'm five years old. But that's just the real world. In the one I make up a little bit more every day, I'm an all-American crime-fighting hero.

Mom's a thousand miles away. On the odd occasion Dad pops out of his bottle of gin, my brother and I have no idea whether he'll be kind and caring Dad or exploding inevitable Dad. Under the influence of unprocessed terror sequestered from moment-to-moment awareness, I generate narcissistic fantasies that serve to occlude from myself and others an agonizing deficit of actual self esteem.

In my virtual reality not only am I a superhero but a remarkably human-like robot. I happily spend hours drawing the circuits and gears and various gizmos that make up my insides. At one point, before Dad tears it down and tosses it, I have a life-size portrait of my exquisitely arranged technological interiors taped to the wall of my room.

Nothing can touch me.

Years later, pacing in the carport after school one day, I try untangling the web of conflicting beliefs my father has instilled in me. The most important thing, he counsels, is freedom. Yet he also tells me all our thoughts and actions are determined by our genes and ultimately by physical law. He teaches not only that communism is the greatest force for evil in the world but also that we're helpless to stanch the red tide as it insidiously defiles civilization. Fate, he calls it, his eyes fixed on some faraway all-consuming power.

I take comfort in the proposition that the invasion of the commies won't change anything fundamental. Why? Because everything we think and feel and choose is pre-determined, every one of us a product of arbitrary design. It was determined that on this afternoon I would walk in circles around an oil stain where the car was usually parked as my next-door neighbor practiced punting a football in the field across the street. Some puppets are made of cotton; others are made of meat.

The question of freedom boils down to the mystery of time. As Bergson points out, time "prevents everything from being given at once."[1] In time there is possibility. Bergson once got himself dragged into an impromptu public debate with Einstein, who thought all of time is laid out from end to end. *Look, over there. It's the future. See? Everything has already happened.*

Einstein famously summed up his view in a letter to a colleague. "For us convinced physicists, the distinction between past, present and future is an illusion, although a persistent one."[2]

An odd statement to be sure, as if Einstein, disbelieving his own pronouncement, felt compelled to undermine it with those last four words. He may as well have said persistence is an illusion, "although a persistent one."

In 1949 Kurt Gödel offered up a cosmological model compatible with travel to the past. Einstein, though he'd helped inspire Gödel's model, dismissed it with the observation that he didn't think he could "telegraph back to his own past."[3] Instead of praise, Gödel got the royal brush off but also a very useful piece of advice: have another look at the phenomenon of irreversibility.

Galileo and Newton incorporated time into their equations at the cost of neutering it. Like many mathematical figures, the t of time can come with a minus sign. When t is positive, Newton's deterministic equations of motion can be utilized to predict the outcome of a given predicament. Where t is negative, the equations allow us to *retrodict* whatever preceded the situation under scrutiny.

For centuries the object of physics has been to step outside the flow of time so as to capture eternal truth about nature. Onboard the mathematics express, Newton and his contemporaries thought they were storming the mind of God, retrieving the principles by which the universe had been crafted. This attitude remained in place right down to Einstein and Schrödinger. Not only relativity but quantum theory is based on deterministic time-reversible equations.

But something funny happened on the way to transcendence. The trouble with perfect knowledge goes back to Rudolf Clausius, who discovered entropy, and Ludwig Boltzmann, who explained it in terms of the irreversible increase of disorder in the aggregate. Clausius introduced the arrow of time to physics in 1865, only six years after the publication of Darwin's *On the Origin of Species*, and called it entropy, roughly the Greek equivalent to the Latin-derived "evolution." Boltzmann's key insight was triggered by Darwin's observation that even though individuals adapt, what evolves is the species to which they belong. So too the irreversible march to uniformity belongs to the gas as a whole regardless of the random motions of the molecules comprising it.

Penrose illustrates the dilemma for theorists well aware that the fundamental laws leave no room for irreversible processes. He seems downright irked at entropy for suggesting, in the midst of all this beautiful two-way deterministic math, that time has a single direction. "What we seem to have deduced is a time-asymmetric law when the underlying physics [is] symmetrical in time."[4]

Physics reduces the present, the *now* of lived experience, to a mathematical figure: t-zero. Starting at this point, the future begins with t-one and proceeds to t-two, etc., while the past starts with t-minus one and carries on from there. Time-symmetry demands that if entropy yields increasing disorder as we project into the future, it must do the same as we project into the past. But as we pass through t-minus one, t-minus two and so on, we find that order increases. "What has gone wrong?" asks Penrose.[5]

What's wrong is that no transcendent law demands backward flowing time so as to cancel out its polar opposite. Time is a one way street. Like myth and religion, physicalist philosophy sacrifices time on the altar of immutable truth.

Let's say you've got a snapshot of a football midair, and along with its location you also have its mass and velocity. By projecting in time, you can predict where the football will land. By projecting backward you can retrodict how hard and in what direction it was kicked in the first place. But a handy methodology doesn't equate to a principle of nature. That we can calculate what preceded the present moment doesn't mean time is literally reversible.

Simple as that observation is, you'd never suspect it from listening to physicists. Richard Feynman went so far as to claim that if particles go forward in time, their complementary "antiparticles" must be stuck in reverse.[6]

In his 1996 offering, *The End of Certainty*, Nobel Prize-winning theorist Ilya Prigogine relegates the time-reversible deterministic universe to the history of failed ideas. Prigogine studied thermodynamics at the Free University of Brussels under the tutelage of Theophile De Donder, who met with hostility from fellow physicists unwilling to grant significance to a merely approximate statistical phenomenon. Prigogine himself was repeatedly told that irreversible processes are merely "transient phenomena" and therefore unworthy of scientific investigation.[7]

Yet we ourselves are transient phenomena. Irreversible processes are implicated in every aspect of our world, from sunlight to weather to oil stains on pavement to the inner workings of our bodies. Irreversibility is associated with evolution as much as entropy, with not only the decrease of order in closed systems but its increase in open systems like us. As

Prigogine writes, "irreversible processes are the rule and reversible processes the exception. Reversible processes correspond to idealizations: we have to ignore the friction to make the pendulum move reversibly."[8]

Newton's model works nicely when applied to a pair of objects such as the sun and the earth, but try adding anything else, and it's dicey at best. Where three or more objects are related in their motion, such as sun, earth and moon, the equations break down.

At the turn of the twentieth century, mathematician Jules-Henri Poincaré discovered that trajectories, when multiple bodies are involved, become coupled in exactly the same way sounds become harmonically coupled in musical notes. Whether in terms of acoustics or dynamics, resonances cannot be pinpointed to exact locations. When confronted by these pesky couplings, scientists can no longer compute the motion of a given system in terms of the trajectories of its individual components. Everything gets mashed up, and none of it yields to the probing eye of physics.[9]

Though many of his peers were happy to dispose of freedom in exchange for certainty, Prigogine welcomed Poincaré's discovery. And why not? Where deterministic equations fail, possibilities erupt.

Poincaré resonances are like army ants systematically eating away at the fabric of determinism, leaving great blotches of instability in their wake. The new dynamic is fitful and imbalanced, out of equilibrium with the environment, oblivious to the deterministic principles to which physically correct systems conform. Eventually our unhinged organism of physics reaches a point where further growth along the same lines is untenable, after which it either disintegrates or evolves to a new level of organization and efficiency. Since choice is fundamental to this process, "bifurcation" cannot be subordinated to deterministic forces. Far from derivative of the randomness of the molecule, freedom is embedded in the marrow of nature's complexity.[10]

Resonance-driven instability generates a slew of interesting phenomena like vortex formation, synchronized chemical oscillations and lasers, in other words, *coherence*. Matter, says Prigogine, takes on new properties in the context of instability, becoming active and organized where otherwise passive and blind.[11] This is the kind of materialism Darwin sought, a materialism that allows creativity to bubble up from the brains of adaptive creatures, in contrast to the sterile and abstract genetic reductionism preferred by his successors.

The myth of universal determinism withers and dies under the glare of nonequilibrium thermodynamics. No equation specifies a whirlpool's opportunistic emergence over a drain. There's no calculating away the spontaneity of a hurricane. The actions of ocean currents and weather systems are unknowable in advance, not just in practice but in principle.

As the body is built up from self-contained mobile units of life called cells, so the cell is founded on nothing more solid than cycling systems of chemical reactions. A multicellular organism is a system of systems, a coherent complexity residing far from the passive equilibrium of the closed-off deterministic world. Our freedom is proportional to our distance from equilibrium. To revert to equilibrium is to be enslaved to physics, i.e. dead.

Like a cicada trapped in its nymph shell unable to take flight, the regressive ego is isolated from the nourishing presence that animates humanity. Though the mechanistic mindset tells us nature forbids true freedom, the only real cage is the one we construct ourselves. Even a physicist is free to reject physicalism, just as a capitalist is free to reject the current antisocial order.

The paradox at the heart of physicalism is how we're devoid of self-determination, yet we must take control over every conceivable facet of existence. When we accept our freedom of thought and therefore responsibility to reason, the first thing we learn is when to let go and let nature do its work, whether outside of us or within.

It's easy to get a blind spot. It's easy to not be aware of what you can't see. Exposed to bad music of the 50s, you instantly know it's pure cheese. But the exact same substance dripping all over and down the sides of pop music six decades later wafts by unnoticed, at least by vast numbers of mutant youth.

This is how it is with everything – with politics, work, consumption, healthcare, food, education, TV… it's all weirdly distorted in a way people generally don't notice because it's always been like that. It's got mnemic homophony behind it, camouflaged against itself over time. What sticks out, what immediately arouses regressive-mutant suspicion, is suddenly making sense where before there was only clueless chatter, a mass of automatons entangled in delusion.

Evolve or collapse applies at some point to every self-organized system, including civilizations. Collapse is the historical default. It's what we get if we follow the morphic brick road predating by eons the rise of modernity. By contrast the first option requires an act of will, specifically the willingness to imagine, to cook up new scenarios in superposition and pursue the best options. Disorganism isn't going to take our hand and lead us to the promised land. We must reflect and reason and plan the world we want to inhabit, or it won't happen. The success of popular opposition to the bombing of Syria signifies a rejection of collective insanity in favor of collective wisdom.

Our task, still ongoing, is to slay the beast self-organizing from the great primordial soup of unreflective thought, leveling societies and ecosystems alike, consuming cultures and spinning off cults like dead skin from a snake. Even now the hum of our ancestors directs us to wake up and return to flowing presence, opening out to creation and co-creating instead of closing off from the source of life so as to subjugate and manipulate, poisoning our mind and our land till our time is up.

Late one Christmas night the commies finally came for my

dad, escorting him to that great police state in the sky. But Dad's freedom was already long since lost. In the timeless and lifeless womb of perfect certainty, nothing will ever touch you.

1. Prigogine, Ilya, *The End of Certainty: Time, Chaos and the Laws of Nature*, New York: Free Press, 1997, p 14
2. Quoted in Prigogine, p 165
3. Quoted in Prigogine, p 165
4. Penrose, Roger, *The Road to Reality*, New York: Knopf, 2005, p 697
5. *Ibid*, p 698
6. *Ibid*, p 639
7. Prigogine, p 62
8. *Ibid*, p 18
9. *Ibid*, p 122
10. *Ibid*, p 44
11. *Ibid*, p 65

Chapter 17

A Dream Called Hell

Who's Peggy?
— Firesign Theater

You know you're in trouble when the clouds turn a greenish shade of black and start spinning on a vortex. Out pops a funnel feeling its way toward something, and before you know it, ground and sky are united in a screaming twisted rope of air. Its purpose? To seek you out. And then? Suck you into the stratosphere. For this is no ordinary tornado. The landscape around you is you. The one-eyed monster has touched down in your mind. Does that mean I'm only imagining it? Not at all. If anything, it's imagining you.

Strange things happen in those few minutes of intense brain activity after lungs and heart have given out. Cardiologist Maurice Rawlings tells the story of a patient, Charlie, whose heart stopped beating in the middle of a pacemaker procedure. Several times Dr Rawlings had to pause during CPR to adjust the location of the device. At some point Charlie piped up, "Don't stop, don't stop. I'm in hell, I'm in hell." When a patient responds during CPR, it's usually to object to the rib-cracking pressure, but not so with Charlie. "For God's sake don't stop. Don't you understand? Every time you stop I'm in hell."[1]

When Charlie begged for a prayer from Rawlings, the good doctor made one up on the spot, and Charlie calmed down after repeating it a few times.

A similar story is told by Howard Storm, an atheist whose "near death experience" turned him into a minister. When a perforation of the stomach hospitalized him in severe pain, Storm seemed to be at the end of his tether. His life had not been a happy one. He says he'd often exploded into rage and intimidated people so as to control them. He couldn't be content with having his own beliefs but had to proselytize his atheism, viewing each encounter with a believer as an opportunity to express his hostility to religion.[2]

So Storm was understandably upset when, after saying goodbye to his wife, he didn't fade to black. Instead he found himself staring down at his body in bed. After enjoying the sensation of clenching his fist in his brand new body, he tried to get his wife's attention, but she didn't notice, so he yelled at her until voices from the hallway began calling him by name, promising to "fix him up" if he followed them.

He stepped out of the room into a fog peopled by shadowy figures that always remained about 15 or 20 feet away. He tried to catch up with them, but they kept receding, telling him with increasing urgency to hurry up and that he'd soon "find out."

The journey, says Storm, went on for hours. The more anxious he got and the more he peppered his entourage with questions, the more aggressive they got in return. They called him pathetic, teasing him over the fact that his hospital gown left his rear end exposed. Worse, the "strange beings" were multiplying into a mob.

> Finally, I told them that I wouldn't go any farther. At that time they changed completely. They became much more aggressive and insisted that I was going with them. A number of them began to push and shove me, and I responded by hitting back at them. A wild orgy of frenzied taunting, screaming and hitting ensued. I fought like a wild man. All the while it was obvious that they were having great fun.

After being torn apart and half eaten, he heard his own voice telling him to pray. But he didn't know any prayers, so he uttered whatever came to mind, such as "The Lord is my shepherd, I shall not want" and "God bless America." It was like spraying the mob with boiling oil.

After pleading for salvation, he saw a distant star that closed in on him until he was surrounded by light, overwhelmed by the love pouring into him. He felt unworthy and told the light he belonged back in the pit. Instead he was subjected to long-buried memories of pain he'd imposed on others, only now it was all his. He saw in particular how badly he'd treated his father, that his view of his dad as villain and himself as victim had been totally misguided.

When he finally awoke in his hospital bed, he was still in agony but grateful to be alive. Previously he would have dismissed the experience as a curiously realistic but meaningless dream, but now he was sure all of it had really happened.

The truth lies somewhere between. Beyond the fact that his emotional release was as real as any in waking life, his dream conveyed significant information about what happens when we reach the end. Far from a simple segue to oblivion, the content of your final experience is conditioned by your treatment of people close to you, the ones you were supposed to love and revere. If you've spent your life beating people down because you're too scared to wrestle your inner demons, there's hell to pay.

All the events in Storm's dream are staples of the near death experience: removal from the body, visions of hell and heaven, meetings with demons and angels and a "life review" in which important events pass "before your eyes." All these elements are also found in the Tibetan Book of the Dead, which elaborates on them in far greater detail than anything in the Western literature.

The cross-cultural similarity of life-ending experience indicates a common ground of memory. Underneath the conscious mind is an unreflective bundle of memory and habit specific to the individual, and underneath the personal unconscious is the collective unconscious. What applies to personal memory applies equally to collective memory. Just as an unusual scent is likely to trigger a recollection of what you were doing the last time it wafted by, an unusual event such as your own imminent death is likely to put you into resonance with previous individuals in the same circumstance. Over thousands of generations, a groove has been laid, and when it's your turn you too are liable to fall into that groove.

The near death experience usually starts with the dreamer departing the body. Some take flight while others stay at the death bed. Some are in darkness while others encounter a brilliant light. Some who find themselves in the dark say it's like a womb, a place of perfect love and contentment, while others get creeped out by how the darkness seems alive. Of those who see a light and get drawn rapidly into its embrace, some luxuriate in it while others burn to a crisp. A golden ring, on closer inspection, can turn out to be a ring of fire.

Regardless of the particulars of your dream, at some point along the way you hit a fork in the groove; you will go one way or the other. If you've bullied and deceived and manipulated your way through life, you're likely to follow those of similarly poor character. Heaven and hell aren't just confabulations or hoaxes but reflections of the structure of the collective unconscious.

The dream of death is a projection of the ego, a movie manufactured by mind in which the viewer somehow continues on as an individual despite being dead. The one thing we can say for sure about death is that the person who once lived no longer exists. The population of death is precisely zero. Yet the idea of death as oblivion is just as hallucinatory as a heaven or hell peopled with the once living.

Part of the allure of materialism is that life is all about making money and acquiring objects of desire while the consequences of greed can safely be ignored because when you die, nothing remains. If you're identical to your body, the end of your body is the end, period. But just because there's no mind without brain doesn't mean the stuff we're made of can be found in our heads.

To think is to juggle abstractions, and built into every abstraction is a degree of vagueness. The concept of a car is a transferal of something tangible into an alien world of fuzziness and uncertainty. A real car exists in a particular moment in time. The abstraction of "car," on the other hand, is a blurred overlapping of countless moments of seeing and driving cars. Where material objects, including the brain, are confined to the current moment, mentality is smeared out from present to distant past.

The clock might say half-past two, but the real time is now. It's always now. Always has been, always will be. We have a fleeting present, which is current only relative to past and future moments, and we have an enduring present completely indifferent to "the time." In this and only this sense, time is an illusion. What's illusory is our commonplace belief, which limits time to its moment-to-moment progression. The reality is not only temporality but absolute presence.

We tend to think of energy in terms of action, neglecting the role of potential energy. Whereas kinetic energy is always spending whatever power it has, potential energy keeps it in reserve. Time is not only the current but the battery.

In the context of implacable presence, temporal flow necessarily yields memory, the covert continuation of the past. What's past in the context of matter is still present in mind. Memory is mental because mentality is the part of us that's rooted directly in absolute presence. Like any material object, the brain is flotsam on the surface of the sea. Only mind, made of time, can plunge into the deep.

Whereas matter is present only to the exclusion of past, mind embraces the past. Rather than one moment replacing the next, the present expands. There's really just one moment, the now, and this moment is breathing us all in. The fleeting materialized moments we mistake for time are only its exhalations.

In space a song is so many sound waves brushing against an ear drum. It's not really a song until we perceive it over time. Only when the past lingers into the present is there melody and rhythm. Without the juxtaposition of the song's past and present, the music dies.

Life is the systematic exploitation of the natural property of memory, every organism a self-contained unit of presence informed by a living past and propelled into an undetermined future.

The earth's biosphere is a branch on the tree of time. There may be other branches as well, but we don't know about them. The universal self-existence of absolute presence is expressed in the local self-existence of microbial life, from which emerges multicellular creatures and the animal kingdom, the vertebrate phylum, the mammalian class, the primate order, the hominid family, the homo genus and humankind.

We exist as individuals in the sense that each of us individuates our species. That the species-mind is unconscious in no way subtracts from its claim on life. What gives us life as individuals is the same animating presence that gives our species life and before that our genus and our family and so on all the way back to the eternal beginning. Death is not really death but only *deindividuation*, a falling back into a deeper form of life, the hive mind, this vast memory pool in which we've been swimming all along, though we had no idea.

Imagine an ocean where all the waves think they're distinct from one another. Yet each wave, after it peaks, reverts to the same mass of water. Though no longer individuated, its core substance carries on.

To remain in the material present, our species requires ongoing individuations. Fear of death is a trick played on us by our species-mind to perpetuate its own life. In reality there's nothing to fear in that darkness, as we merely exchange one form of life for another. Death is life by other means.

Among the most common motifs of the near death experience is the act of passing through a tunnel toward a light. It's easy to see why this image has been abstracted into the species-mind, given that the vast majority of us embark on a journey just like this on the way to birth. In the context of the great unconscious, birth itself is a kind of dying.

If there's a true death, a polar opposite to life, it's the act of stepping out from the universal flow into a stagnant pool. This occurs nowhere in living creation except the human mind. At the level of unconscious thought, sensory and cognitive data are processed without hindrance. With the emergence of conscious reflection and discrimination, however, comes the ability to stifle the flow of knowledge when it doesn't suit our preferences.

After a painful experience — the kind that tells you something you didn't want to know about yourself — you're liable to avoid thinking about it, to blot it out as if it never took place. It's like eating a piece of fruit, and instead of being digested, the fruit is sequestered from the gustatory flow and begins rotting. Pretty soon it gets moldy, starts stinking the place up. People wonder why you have bad breath. Someone who really cares about you sits you down for a talk. *Okay, I get it. I'm stuck on this one issue, and I need to dredge it up and deal with it.* Finally the fruit starts to break down and process. Once more you're in the flow.

Small children are especially adept at concealing inconvenient facts about their behavior, and one of the chief responsibilities of parents is to help their offspring face up to bad things they've done while emphasizing that they're still worthy of love. In this way children integrate both the negative and positive aspects of themselves into their self-understanding.

But some people never learn to accept themselves for what they are, never surpassing a primitive stage of emotional development. They never learn to digest the fruit, and the moldier it gets, the more awful the task of confronting it. What regressives fear is not so much dying as re-entering the dance. In other words they fear life, as they've entombed themselves in ego. Habitually burying memories of their harmful actions, regressives are likely to go on the attack when the correct emotion is regret or sorrow, and they exploit and control others instead of empathizing and cooperating for everyone's good.

Hell is a kind of spiritual vomiting. What comes up is only what you yourself already swallowed but never digested. Energy must be expended to keep it all down, and at the end of your life none remains. Far from being imposed by an all-knowing deity, your hell is already being prepared for you in your personal unconscious.

If we're fundamentally separate and distinct from other people, we never have to face the consequences of our actions. Yet death is the unraveling of precisely this misconception. You find out you weren't exactly you after all but only an individualized expression of the totality of the human race. We're all one. If you've lived your life victimizing people, a shock awaits you when you discover that your victims were just different expressions of your true self.

Like habits clinging to an organism, those you harm stay with you. This is the meaning of *haunted*.

When he was diagnosed with terminal cancer, Republican strategist Lee Atwater was shaken to his core. His maxim, "Never kick a man when he's up," came back to haunt him as he was beset with anguish and shame over his crimes against democracy. It seemed that Atwater, a very capable musician, still had "soul" even after all that deception and back-stabbing in pursuit of power. By confronting himself prior to his final moments and accepting the consequences of his actions, he evaded the compressed self-examination known as hell. The bad thing isn't the sorrow but putting it off until you're locked in a nightmare you'll never return from.

Though united in species-mind, we live out our days in the

plural. *Love* is what you get when you separate the inseparable. Touch reminds us of our singularity. We all need that recollection once in awhile.

When he was 12 years old, Matthew Dovel absorbed an overpowering dose of love during a near death experience. At the end of an exhausting swim, his friends dunked him back under just as he'd surfaced to inhale. Instantly he was out-of-body. In place of the usual parade of thoughts and feelings, there was only white light, intense in a way that can't be fathomed. Memories of good times came flooding in from his living history.[3]

But his vision of paradise was cut short when Jesus appeared out of nowhere, grabbed him by the arm and told him he had to go back. By this point his friends had carried him out of the pool and resuscitated him. How could they have done this to him?

Dovel couldn't shake off the bitterness of being snatched from heaven. Whenever the rage threatened to surface, he kept his brain busy with alcohol and cocaine. He grew up, married and procreated but never learned to look inside and process that disappointment.

So he decided to kill himself. He'd simply swallow a few hundred sleeping pills with a bottle of vodka, and heaven would be his once more. Not surprisingly, the second go-round was different. In free fall over an all-consuming pit, his private timeline reeled in once more except that now the memories weren't so sweet. Screaming in his face, friends returned the pain he'd caused them in his years of sullen selfishness. He saw his young daughter as a teenager contemplating suicide as a consequence of his insane action. Worst of all, he'd have no chance to go back and redeem himself.

Dovel didn't know how to wake up except by going to sleep. To immerse himself in his memory pool was to risk drowning in it. By chance his poison failed to dispatch him, and he walked away from the brink once again — only now a free man.

Safer methods of awakening are available. The simplest route to the deep is meditation, the art of relaxing while remaining alert. You're in a quiet place, unmoving aside from the back-and-forth of breath, but you're also sitting up straight and focused on immediate flowing existence. Pretty soon you settle in; the moment starts stretching out. Time feels good piling up in a single scoop of consciousness. The body senses itself differently, perhaps because you're aware of it over time and not just over space, resonating with the you that commenced the silence and all the you in between then and now. Even the littlest adjustment in posture, a slight arching of the lower back, can radically improve the morphic reception.

What seems like a tiny unobtrusive piece of each of us is in fact the center of us all. There's a place we can go that knows nothing of addiction or fear or resentment or rage, and the more we go there the easier it is to keep going there. All the times you've ever sat in meditation help power that sense of overflowing calm, the peace that echoes off itself as joy. Even those who sat long before your birth are present in you, implicit in your posture, explicated with every breath.

In his near-death primer, *Lessons from the Light*, Kenneth Ring introduces his readers to an accidental messenger named Peggy Holladay, who suffered a compound skull fracture from a car crash. She doesn't recall any profound experience in the immediate aftermath, but years later riding in a van much like the one that crashed, her stopper came undone, her individuated psyche swirling over the drain.[4]

In tears Peggy gets up from the passenger seat and heads to the back of the van to lie down. Even against the current unleashed by her pounding heart, she manages to slow down and deepen her breath. Eventually she finds herself floating in a womblike darkness she later calls the "blackest blackness." Out of this void comes a light, a pinprick that rapidly expands into a dazzling blue-white field. Despite its intensity, she feels no pain looking into it, and she gets that it's both alive and incomprehensibly intelligent. "I knew the light was a being. I also knew that the light being was God and was genderless."

This makes sense. After all, no one is as brilliant as the collective intelligence from which we individuate. The archetypal light is an image built up over time as people come face to face with the shared form out of which we all originate and into which we eventually return. No gender can be assigned to the unitary mind that gives rise to both female and male.

"I began to sense the light knew me very well right before it surrounded me completely." Well, of course. How could an ocean not know its waves? Immersed in light, she recognized it as "pure love" given to "everyone equally and without any conditions."

In the course of her motorized magic carpet ride, Holladay entered a state of grace, that paradoxical place where species-mind coincides with individuated perception. From there she gained access to the collective wisdom of the human race. She learned, for instance, that the world is an illusion or at least that it's real to a lesser degree than it seems. A subtle insight, since material existence has no intrinsic nature but only piggybacks on absolute presence, spatializing and materializing its implicit forms. All this is real, yes, but not *really* real.

CS Lewis suggested that purgatory, rather than a unique place distinct from heaven and hell, is merely hell with a time limit. Time's up when the soul voluntarily disavows its rage against God, and the flames that once scorched now purify. Lewis even took it a step further by proposing that hell is only a state of mind while heaven is "reality itself." Rather than two possible destinations there's only heaven, but those who are lost perceive it as hell.[5]

A messenger at the front lines in World War I, Adolf Hitler frequently had to cross enemy lines. Though he always found his way back, he was lost in a different way. He survived every soldier in his company — all shot down or blown up through the miracle of advanced technology — but he had no notion that pain is something to be processed, brought to the surface and expelled. He could only replicate, blindly, the mechanized killing of that war with the holocaust.

In the unconscious pits of his mind, Hitler knew perfectly well he was a mass murderer, but he couldn't face up to it. He preferred to see himself as savior of Germany. His 1939 invasion of Poland was only to provide a buffer against a Soviet incursion which, in his paranoia, he believed to be inevitable. All those Jews, Roma, Slavs, gays and socialists had to be eliminated because their intrinsically subversive nature threatened the fatherland and therefore civilization. Nobody played the role of gallant defender better than Adolf. But in the course of pursuing his fantasy, a schism developed between the unconscious and conscious parts of his mind. A gap, or gradient, emerged between what he thought of himself and what he knew deep down.

Nature abhors gradients. It seeks to wipe them out. Hitler could stave off the elimination of his mental gradient only for as long as he kept up his strength. The schism can't survive the individual who cultivates it. Private delusion washes away as body-mind tips over and spills into species-mind.

Instead of pausing to reflect on his crimes, Adolf kept ordering more. Whether his own troops or associates he perceived as threats or anyone fitting his idea of subhuman, each victim was another serving down the hatch, added to the already fragrant pile. We're talking upwards of 50 million ex-people. That's a lot of undigested organic matter.

When the going was good, Adolf commanded his architect Albert Speer to draw up plans for a sort of Nazi Vatican to dominate the new Berlin. It never got off the ground. By the spring of 1945, says Speer, Hitler was "shriveling up like an old man. His limbs trembled; he walked stooped, with dragging footsteps." His formerly masterful speech gave way to a "faltering, toneless manner of speaking. When he became excited, as he frequently did in a senile way, his voice would start breaking."

Well, naturally. At his unconscious core, Hitler knew what was to come. No wonder his "complexion was sallow, his face swollen; his uniform, which in the past he had kept scrupulously neat, was often neglected in this last period of life and stained by food he had eaten with a shaking hand."[6]

As the world caved in he took careful aim at his life and swatted it with a quick and irreversible motion. Like a tornado wiping out the gradient between warm air on the ground and cold air in the sky, all the emotional processing he failed to undergo in his life contracted into a singularity of immeasurable agony. Too cowardly to accept what he did to even one person, he now faced the bottomless horror of millions. In effect, all their suffering was his.

The bullet, already sailing from the barrel of his firmly gripped apparatus, ripped through his brain in fractions of a second. But time is funny. If vast processing is required, the moment dilates beyond recognition.

Hitler went to hell and goes to hell and can't stop going to hell until every last pocket of delusion has emptied out. There's a reason hell is said to last forever. It doesn't really of course, as everything re-enters the flow. The good news for the Adolfs of the world is that hell ends rather than beginning with the termination of individuated life. The bad news? Eternity, if need be, will squeeze itself into a microsecond.

You're walking alone in a black and white forest, the air wet and heavy and devoid of activity.

Silence.

But it's not the eerie quiet broken only by a distant rising wind that concerns you. As a matter of fact, it's that tree over there. No, not that one — the little one all knotted up. You've seen it before, possibly many times and always from exactly the same angle as if you've done nothing but cover the same ground over and over by exactly the same route. If you had an axe you'd go to town on it, hacking away at its corpse until it's a million pieces of pulp. A smile momentarily stretches your cracked lips.

Then a breeze picks up, a slight stirring from above, and you remember.

1. Rawlings, Maurice, *To Hell and Back*, Nashville: Thomas Nelson, 1993, p 37
2. Storm, Howard, *My Descent Into Death: A Second Chance at Life*, New York: Doubleday, 2005, pp 10-37
3. Johnson, Sylvia and Wallace, Rob, "Touching Heaven and Hell," *20/20*, ABC broadcast July 10, 2007 Available from: http://abcnews.go.com/2020/story?id=3359251&page=1#.Tzlc5YHpiSo (Accessed Jan 4 2014)
4. Ring, Kenneth and Valarino, Evelyn, *Lessons from the Light*, Needham, Massachusetts: Moment Point Press, 2006, pp 42-49
5. Lewis, CS, *The Great Divorce*, New York: HarperCollins, 2001 (1946) pp 68-70
6. Speer, Albert, *Inside the Third Reich*, New York: Simon & Schuster, p 472

Acknowledgements

Many people helped make this book a reality, whether through enlightening me, publishing my work, commenting on my drafts, suggesting other sources of information, catalyzing insights during conversation or simply handing me a book to read. These include my original teacher, Wallace Dace, as well as University of Kansas professors Alfonso Verdu, Franklin Shontz, Ted Johnson and Cheryl Lester. A big thank you goes to Bruce Anderson, not just for all the articles he kindly published but for all the ones he wisely didn't. Thanks also to Csaba Polony, Anthony Freeman, Huping Hu, Lydia Sargent, Gilles d'Amery, Rupert Sheldrake, Miroslav Hill, Mark Scaramella, George Frazier, Richard C White, Whitney Hodgin, Liam Sprod, Maryann Baum, Alex Shahverdiloo, Jefferson Listerman, Patrick Lawson, Scott Chase, Derek Gatherer, Melissa Parson, Bharath Narayanan, The McCallum's, The Mathematics, Dustin Seabourn, Chase Dahnke, Miriam Climenhaga and Royce Kitts, among many others.

And finally, thanks Mom, you made it all possible.

Credits

Chapters 1, 2, 8, 10, 11, 16 and 17 are adapted from articles originally published in the *Anderson Valley Advertiser*. Some of these articles reappeared, with revisions, on the website, *Swans*. An early version of Chapter 2 also appeared in *Prespacetime Journal*. Chapters 4, 5 and 6 are adapted from articles originally published on the website *Skeptical Investigations*. An early version of Chapter 4 also appeared in the *Journal of Consciousness Studies*. Chapters 7, 9, 12 and 15 are adapted from articles originally published in *Left Curve*. A section of Chapter 14 first appeared on the websites *Counterpunch* and *Amazon*.

BOOKS

Iff Books is interested in ideas and reasoning. It publishes material on science, philosophy and law. Iff Books aims to work with authors and titles that augment our understanding of the human condition, society and civilisation, and the world or universe in which we live.